Algebras and Lattice in Hawai'i
a conference in honor of
Ralph Freese, William Lampe, and J.B. Nation

Preface

It is with great pleasure that we present the Proceedings of ALHawai'i 2018: Algebras and Lattices in Hawaii: honoring Ralph Freese, Bill Lampe and JB Nation. The variety of papers presented here represents a small sampling of the vast influence that these men have had on the Universal Algebra and Lattice Theory community. Their combined contributions to publications is over 170 papers and books.

Ralph and Bill joined the University of Hawai'i Mathematics Department in 1972. Ralph had just completed his PhD under Robert Dilworth at the California Institute of Technology. Bill had completed his PhD in 1969 at Pennsylvania State University under Orrin Frink and George Grätzer. In 1979 JB joined the Department. He had also worked under Dilworth at Caltech, earning his doctorate there in 1973.

Those who have been fortunate enough to go to Hawaii and meet with Freese, Lampe, and Nation have known true hospitality. A greeting at the airport with a lei from Bill, a dinner party in Ralph's backyard paradise, or a trumpet serenade from JB are all likely events.

We, the planning committee, cannot thank these men enough for the time they have spent with us doing mathematics. These collaborations have led to many fruitful results. Freese, Lampe, and Nation continue to be active and produce remarkable work. We are grateful for the opportunity to honor these three great mathematicians, and we are happy that many conference participants could join us for three days of celebration.

Kira Adaricheva
William DeMeo
Jennifer Hyndman

21 April 2018

Organizing Committee

Kira Adaricheva	Hofstra University, Hempstead
William DeMeo	University of Colorado, Boulder
Jennifer Hyndman	Univ. Northern British Columbia, Prince George

Referees

Kira Adaricheva	Hofstra University, Hempstead
Charlotte Aten	University of Rochester, Rochester
Clifford Bergman	Iowa State University, Ames
David Casperson	Univ. Northern British Columbia, Prince George
William DeMeo	University of Colorado, Boulder
Nikolaos Galatos	University of Denver, Denver
Jerrold Griggs	University of South Carolina, Columbia
Paweł Idziak	Jagiellonian University, Krakow
Peter Jipsen	Chapman University, Orange
Peter Jones	Marquette University, Milwaukee
Keith Kearnes	University of Colorado, Boulder
Roger Maddux	Iowa State University, Ames
Peter Mayr	University of Colorado, Boulder
George McNulty	University of South Carolina, Columbia
Péter P. Pálfy	Alfréd Rényi Institute of Mathematics, Budapest
Agata Pilitowska	Politechnika Warszawska, Warsaw
Robert Proctor	University of North Carolina, Chapel Hill
Francesco Sica	Nazarbayev University, Astana
Sylvia Silberger	Hofstra University, Hempstead
David Stanovsky	Charles University, Prague
Matt Valeriote	McMaster University, Hamilton
Anna Zamojska-Dzienio	Politechnika Warszawska, Warsaw

Table of Contents

Universality and Q-universality in varieties of quasi-Stone algebras

M. E. Adams, W. Dziobiak, and H. P. Sankappanavar

[1] Department of Mathematics, State University of New York, New Paltz 12561
adamsm@newpaltz.edu
[2] Department of Mathematics, University of Puerto Rico, Mayagüez 00681
w.dziobiak@gmail.com
[3] Department of Mathematics, State University of New York, New Paltz 12561
sankapph@newpaltz.edu

Keywords: variety of algebras, quasivariety of algebras, critical algebra, universal variety, Q-universal variety, identity, quasi-Stone algebra, Priestley duality

Abstract. The lattice $L_V(\mathbf{QS})$ of subvarieties of the variety $\mathbf{QS}$ of quasi-Stone algebras ordered by inclusion is an $\omega + 1$ chain.
It is shown that the lattice $L_Q(\mathbf{Q_{0,1}})$ of subquasivarieties of the variety $\mathbf{Q_{0,1}}$ is a 4-element chain (where $\mathbf{Q_{0,1}}$ is the variety of height 3 in $L_V(\mathbf{QS})$), $L_Q(\mathbf{Q_{2,0}})$ is a finite non-modular lattice (where $\mathbf{Q_{2,0}}$ is the variety of height 4), $L_Q(\mathbf{Q_{3,0}})$ is still a finite lattice (where $\mathbf{Q_{3,0}}$ is the variety of height 7), whilst $L_Q(\mathbf{Q_{2,1}})$ is a countably infinite lattice of finite breadth, thereby satisfying a non-trivial lattice identity, and is locally finite (where $\mathbf{Q_{2,1}}$ is the variety of height 8). In the process, the critical algebras in $\mathbf{Q_{2,1}}$ are completely determined.
It is further shown that $L_Q(\mathbf{Q_{1,2}})$ is finite-to-finite relatively universal (in the sense of Hedrlín and Pultr), hence, it is uncountable and does not have finite breadth (where $\mathbf{Q_{1,2}}$ is the variety of height 9). Furthermore, it is shown that $L_Q(\mathbf{Q_{1,2}})$ is not Q-universal (in the sense of Sapir), thereby showing false a long-standing conjecture that every finite-to-finite relatively universal variety is Q-universal.
Finally, it is shown that the variety $\mathbf{Q_{2,2}}$ (of height 13) is finite-to-finite universal and, hence, Q-universal. It follows, for example, that the lattice $L_Q(\mathbf{Q_{2,2}})$ has a free lattice on a countably infinite set of generators as a sublattice (thereby failing every non-trivial lattice identity).
No proper subvariety of $\mathbf{Q_{1,2}}$ is finite-to-finite relatively universal to any of its proper subvarieties, nor is any proper subvariety of $\mathbf{Q_{2,2}}$ universal.

1   Quasi-Stone algebras

As introduced in [32], an algebra $L = (L; \vee, \wedge,', 0, 1)$ of type $(2, 2, 1, 0, 0)$ is a *quasi-Stone algebra* if

(i) $(L; \vee, \wedge, 0, 1)$ is a bounded distributive lattice;
(ii) $0' = 1$ and $1' = 0$;
(iii) $(x \vee y)' = x' \wedge y'$;
(iv) $(x \wedge y')' = x' \vee y''$;
(v) $x \le x''$;
(vi) $x' \vee x'' = 1$.

For $m \in \omega$, B_m denotes the Boolean lattice with m atoms and $\widehat{B}_m$ denotes the lattice $B_m \oplus \{1_m\}$ where 1_m is a new element and $\oplus$ denotes the ordinal sum. For $m, n \in \omega$, $Q_{m,n}$ denotes the quasi-Stone algebra $(\widehat{B}_m \times B_n; \vee, \wedge,', (0, 0), (1_m, 1))$ where

$$(x, y)' = \begin{cases} (0, 0) & \text{if } (x, y) \ne (0, 0), \\ (1_m, 1) & \text{otherwise.} \end{cases}$$

As shown in [32], the lattice of varieties of quasi-Stone algebras $L_V(\mathbf{QS})$ under inclusion forms an $\omega + 1$ chain

$\mathbf{T} \subset$
$\mathbf{Q_{0,0}} \subset$
$\mathbf{Q_{1,0}} \subset \mathbf{Q_{0,1}} \subset$
$\mathbf{Q_{2,0}} \subset \mathbf{Q_{1,1}} \subset \mathbf{Q_{0,2}} \subset$
$\mathbf{Q_{3,0}} \subset \mathbf{Q_{2,1}} \subset \mathbf{Q_{1,2}} \subset \mathbf{Q_{0,3}} \subset$
$\mathbf{Q_{4,0}} \subset \mathbf{Q_{3,1}} \subset \mathbf{Q_{2,2}} \subset \mathbf{Q_{1,3}} \subset \mathbf{Q_{0,4}} \subset \ \ldots \ldots \ \subset \mathbf{QS},$

where $\mathbf{T}$, $\mathbf{Q_{m,n}}$, and $\mathbf{QS}$ denote the trivial variety, the variety $\mathbf{V}(Q_{m,n})$ of quasi-Stone algebras generated by $Q_{m,n}$ ($m, n \in \omega$), and the variety of all quasi-Stone algebras, respectively.

2   Quasivarieties of quasi-Stone algebras

For a variety $\mathbf{V}$, let $L_Q(\mathbf{V})$ denote the lattice of subquasivarieties of $\mathbf{V}$ ordered by inclusion. Then, in particular, $L_Q(\mathbf{V})$ may be regarded as a measure of the complexity of the variety $\mathbf{V}$. The following holds:

Theorem 1. *For the variety of quasi-Stone algebras* **QS***:*
  (i) $L_Q(\mathbf{Q}_{0,1})$ *is a 5-element chain;*
  (ii) $L_Q(\mathbf{Q}_{2,0})$ *is a 12-element non-modular lattice;*
  (iii) $L_Q(\mathbf{Q}_{3,0})$ *is a finite lattice;*
  (iv) $L_Q(\mathbf{Q}_{2,1})$ *is a countably infinite lattice which is locally finite and has finite breadth.*

With respect to 1 (ii), note that Gorbunov [13] showed that if any lattice of subquasivarieties is modular, then it is distributive.

A finite algebra *A* is *critical* if it is not embeddable in a direct product of its proper subalgebras, that is,

$$A \notin \mathbf{ISP}(\{B : B \text{ is a proper subalgebra of } A\}).$$

It is well known and not difficult to prove that a locally finite quasivariety is determined by its critical algebras. As shown in [32], the variety of quasi-Stone algebras **QS** is locally finite. Thus, identifying all the critical algebras in a variety **V** of quasi-Stone algebras enables one to determine, at least theoretically, $L_Q(\mathbf{V})$. In particular, 1 is established by determining the critical algebras contained in each variety **V** of quasi-Stone algebras for $\mathbf{V} \subseteq \mathbf{Q}_{2,1}$.

That there is a significant difference between the variety $\mathbf{Q}_{2,1}$ and the variety $\mathbf{Q}_{1,2}$, where critical algebras are not readily identified, is suggested by the following:

Theorem 2. *For the variety of quasi-Stone algebras* $\mathbf{Q}_{1,2}$*,* $L_Q(\mathbf{Q}_{1,2})$ *is uncountable.*

In fact 2 is an immediate consequence of 4. However, before considering $L_Q(\mathbf{Q}_{1,2})$ further, we jump to the variety of quasi-Stone algebras $\mathbf{Q}_{2,2}$. First, we give some terminology.

A *graph* $G = (V, E)$ is a set of *vertices* V and a set of *edges* E of 2-element subsets of V. For graphs $G = (V, E)$ and $H = (W, F)$, a mapping $\phi: G \to H$ is *compatible* providing, for every $\{x, y\} \in E$, $\{\phi(x), \phi(y)\} \in F$. A variety **V** is *universal* providing every category of algebras of finite type is isomorphic to a full subcategory of **V**, equivalently, the category **G** of all graphs and all compatible mappings is isomorphic to a full subcategory of **V**, see Hedrlín and Pultr [16] (as well as Pultr and Trnková [31]). If there exists a functor $\Phi: \mathbf{G} \to \mathbf{V}$ which establishes that **V** is universal and, in addition, sends finite graphs to finite algebras, then **V** is said to be *finite-to-finite universal*.

Theorem 3. $Q_{2,2}$ *is finite-to-finite universal.*

For a quasi-Stone algebra L, let $\text{End}(L)$ denote the monoid of endo-morphisms of L under composition. The following is an immediate consequence of 3 together with known properties of graphs.

Corollary 1. *For every monoid M and cardinal $\kappa \geq |M| + \omega$, there exists a family of quasi-Stone algebras $(L_i : i \in I)$ such that*
   (i) $L_i \in Q_{2,2}$ for $i \in I$,
   (ii) $L_i \not\cong L_j$ for distinct $i, j \in I$,
   (iii) $\text{End}(L_i) \cong M$ for $i \in I$,
   (iv) $|I| = 2^\kappa$ and $|L_i| = \kappa$ for $i \in I$.
Moreover, if $|M|$ is finite, then there also exists a countably infinite family of finite quasi-Stone algebras $(L_i : i \in I)$ satisfying (i), (ii), and (iii).

Furthermore, 3 is sharp in the sense that the largest proper subvariety of $Q_{2,2}$, namely $Q_{3,1}$, is not universal (since, up to isomorphism, there are only two algebras in $Q_{3,1}$ with a trivial endomorphim monoid).

Another measure of the complexity of a variety V is a notion of Q-universality, as introduced by Sapir. A variety V of algebras of finite type is Q-*universal* providing that, for any quasivariety W of finite type, $L_Q(W)$ is a homomorphic image of a sublattice of $L_Q(V)$. This notion was introduced by Sapir [33] where, amongst other results, he showed that the variety of commutative 3-nilpotent semigroups is Q-universal.

As shown in [2], every finite-to-finite universal variety V has the ideal lattice of a free lattice on countably many generators as a sublattice of $L_Q(V)$ and, hence, is Q-universal. In particular, the following is an immediate consequence of 3:

Corollary 2. $Q_{2,2}$ *is Q-universal.*

We now return to the variety of quasi-Stone algebras $Q_{1,2}$.

Even though a variety may not be universal, it may fail to be universal for superficial reasons. Various attempts to capture this phenomenon may be found in the literature (see, for example, Demlová and Koubek [8], [9] or Sichler [36]). One such notion, introduced in this context, is that of *relative universality*. A variety V is *relatively universal* to a proper subvariety W (or, briefly, W-*universal*) if there is a faithful functor $\Psi\colon G \to V$ such that $\text{Im}(\Psi(f))$ belongs to W for no compatible mapping f and if

$\phi : \Psi(G) \to \Psi(G')$ is a homomorphism, where G and G' are graphs, then either $\text{Im}(\phi)$ belongs to $\mathbf{W}$ or $\phi = \Psi(f)$ for a compatible mapping $f : G \to G'$. If, in addition, Ψ assigns finite algebras to finite graphs, $\mathbf{V}$ is said to be *finite-to-finite* $\mathbf{W}$-*universal*.

Theorem 4. $\mathbf{Q}_{1,2}$ *is finite-to-finite* $\mathbf{Q}_{2,1}$-*universal*.

Since $\mathbf{Q}_{1,2}$ is finite-to-finite relatively universal, it follows immediately that $L_Q(\mathbf{Q}_{1,2})$ is uncountable (see 2). In turn, $\mathbf{Q}_{2,1}$ is not finite-to-finite universal relative to any proper subvariety, since, by 1 (iv), $L_Q(\mathbf{Q}_{2,1})$ is countable, showing that 4 is sharp in this context.

From the outset, the question was raised as to how the hypothesis that a variety be finite-to-finite universal could be weakened and still conclude that it be Q-universal (see Question 20 [1]). At the time we had conjectured that finite-to-finite relative universality should be sufficient to give Q-universality (see Question 22 [1]).

A number of papers have considered universality, weaker notions of universality, as well as Q-universality in different combinations and in different contexts: for example, Basheyeva, Nurakunov, Schwidefsky, and Zamojska-Dzienio [6], Demlová and Koubek [8,9,10,11], Koubek and Sichler [17,18,19,20,21,22,23], Kravchenko [24,25,26,27,28], Nurakunov, Semenova and Zamojska-Dzienio [29], Schwidefsky and Zamojska-Dzienio [34], Semenova and Zamojska-Dzienio [35]. Nevertheless, whether finite-to-finite relative universality was sufficient to imply Q-universality has proven evasive. This longstanding and quietly irksome question explains our particular interest in the variety $\mathbf{Q}_{1,2}$:

Theorem 5. *The variety* $\mathbf{Q}_{1,2}$ *is finite-to-finite relatively universal, but not Q-universal.*

3 Concluding remarks

All proofs rely heavily on Gaitán's [12] version of Priestley duality for quasi-Stone algebras (cf. Cignoli [7], Halmos [15], and [3].)

We found the proof of 5 to be very tricky and over an extended period of time we threw everything at it including the kitchen sink. As a consequence, we are left with a number of related questions. For example, because $L_Q(\mathbf{Q}_{2,1})$ has finite breadth, it satisfies a nontrivial lattice identity. We also know that there is a homomorphism f from $L_Q(\mathbf{Q}_{1,2})$ onto $L_Q(\mathbf{Q}_{2,1})$ such that every congruence class of ker f is a distributive lattice. We ask whether, in general, given a lattice L and a congruence Θ on L such that $[a]\Theta$ is distributive for every $a \in L$ and L/Θ satisfies a non-trivial lattice identity, does it follow that L satisfies a non-trivial lattice identity?

Details of all of the above are found in [4] and [5].

References

1. M.E.Adams, K.V.Adaricheva, W.Dziobiak, and A.V.Kravchenko, *Some open questions related to the problem of Birkhoff and Maltsev*, Stud. Logic 78 (2004), 357–378. 5

2. M.E.Adams and W.Dziobiak, *Finite-to-finite universal quasivarieties are Q-universal*, Algebra Universalis 46 (2001), 253–283. 4

3. M.E.Adams and W.Dziobiak, *Endomorphisms of distributive lattices with a quantifier*, Internat. J. Algebra Comput. 17 (2007), 1349–1376. 5

4. M.E.Adams, W.Dziobiak, and H.P.Sankappanavar, *Universal varieties of quasi-Stone algebras*, Algebra Universalis 76 (2016), 155–182. 6

5. M.E.Adams, W.Dziobiak, and H.P.Sankappanavar, *Quasivarieties of quasi-Stone algebras*, preprint. 6

6. A.Basheyeva, A.Nurakunov, M.Schwidefsky, and A.Zamojska-Dzienio, *Lattices of subclasses. III*, Sibirsk. Élektron. Mat. Izv. 14 (2017), 252-263. 5

7. R.Cignoli, *Quantifiers on distributive lattices*, Discrete Math. 96 (1991), 183–197. 5

8. M.Demlová and V.Koubek, *Endomorphism monoids in varieties of bands*, Acta Sci. Math. (Szeged) 66 (2000), 477–516. 4, 5

9. M.Demlová and V.Koubek, *Weaker universality in semigroup varieties*, Novi Sad J. Math. 34 (2004), 37–86. 4, 5

10. M.Demlová and V.Koubek, *Weak alg-universality and Q-universality in semigroup quasivarieties*, Comment. Math. Univ. Carolin. 46 (2005), 257-279. 5

11. M.Demlová and V.Koubek, *On universality of semigroup varieties*, Arch. Math. (Brno) 42 (2006), 357-386. 5

12. H.Gaitán, *Priestley duality for quasi-Stone algebras*, Stud. Logica 64 (2000), 83-92. 5

13. V.A.Gorbunov, *Lattices of quasivarieties*, Algebra and Logic 15 (1976), 275–288. 3

14. V.A.Gorbunov, "Algebraic Theory of Quasivarieties," Plenum Publishing Co., New York, 1998.

15. P.R.Halmos, *Algebraic logic, I. Monadic Boolean algebras*, Compositio Math. 12 (1955), 217–249. 5

16. Z.Hedrlín and A.Pultr, *On full embeddings of categories of algebras*, Illinois J. Math. 10 (1966), 392–406. 3

17. V.Koubek and J.Sichler, *Almost ff-universal and Q-universal varieties of modular 0-lattices*, Colloq. Math. 101 (2004), 161-182. 5

18. V.Koubek and J.Sichler, *On relative universality and Q-universality*, Stud. Logic 78 (2004), 279-291. 5

19. V.Koubek and J.Sichler, *On subquasivarieties of finitely generated varieties of distributive double p-algebras*, Contributions to general algebra. 17, 129143, Heyn, Klagenfurt, 2006. 5

20. V.Koubek and J.Sichler, *On finitely generated varieties of distributive double p-algebras and their subquasivarieties*, Topics in discrete mathematics, 71-92, Algorithms Combin. 26, Springer, Berlin, 2006. 5

21. V.Koubek and J.Sichler, *On synchronized relatively full embeddings and Q-universality*, Cah. Topol. Géom. Différ. Catég. 49 (2008), 289-306. 5

22. V.Koubek and J.Sichler, *Almost ff-universality implies Q-universality*, Appl. Categ. Structures 17 (2009), 419–434. 5

23. V.Koubek and J.Sichler, *On relative universality and Q-universality of finitely generated varieties of Heyting algebras*, Sci. Math. Jpn. 74 (2011), 63-115. 5

24. A.V.Kravchenko, *Q-universal quasivarieties of graphs*, Algebra and Logic 41 (2002), 173–181. 5

25. A.V.Kravchenko, *The complexity of quasivariety lattices for varieties of unary algebras*, Siberian Adv. Math. 12 (2002), 63–76. 5

26. A.V.Kravchenko, *Complexity of quasivariety lattices for varieties of differential groupoids*, Siberian Adv. Math. 19 (2009), 162-171. 5

27. A.V.Kravchenko, *Complexity of lattices of quasivarieties for varieties of differential groupoids. II*, Siberian Adv. Math. 23 (2013), 84-90. 5

28. A.V.Kravchenko, *On the complexity of lattices of quasivarieties for varieties of unary algebras. II*, (Russian) Sib. Èlektron. Mat. Izv. 13 (2016), 388-394. 5

29. A.Nurakunov, M.Semenova and A.Zamojska-Dzienio, *On lattices connected with various types of classes of algebraic structures*, Uch. Zap. Kazan. Univ. Ser. Fiz.-Mat. Nauki. 154 (2012), 167-179. 5

30. H.A.Priestley, *Representation of distributive lattices by means of ordered Stone spaces*, Bull. London Math. Soc. 2 (1970), 186–190.

31. A.Pultr and V.Trnková, "Combinatorial, Algebraic and Topological Representations of Groups, Semigroups and Categories," North-Holland, Amsterdam, 1980. 3

32. N.H.Sankappanavar and H.P.Sankappanavar, *Quasi-Stone algebras*, Math. Logic Quart. 39 (1993), 255–268. 2, 3

33. M.V.Sapir, *The lattice of quasivarieties of semigroups*, Algebra Universalis 21 (1985), 172–180. 4

34. M.Schwidefsky and A.Zamojska-Dzienio, *Lattices of subclasses. II*, Internat. J. Algebra Comput. 24 (2014), 1099-1126. 5

35. M.V.Semenova and A.Zamojska-Dzienio, *On lattices of subclasses*, Sib. Math. J. 53 (2012), 889-905. 5

36. J.Sichler, *Nonconstant endomorphisms of lattices*, Proc. Amer. Math. Soc. 34 (1972), 67–70. 4

Random Relation Algebras

Jeremy F. Alm*

Department of Mathematics, Lamar University, Beaumont, TX 77710
alm.academic@gmail.com

What does a "typical" finite relation algebra look like? In graph theory, one has the "random graph" $G_{n,p}$, which is actually a probability space of graphs [3]. (If one sets $p = \frac{1}{2}$, $G_{n,p}$ corresponds to the uniform distribution on the set of all labelled graphs on n vertices.) Then a graph property P (like being connected) is said to hold in "most" graphs if the probability that P holds in $G_{n,p}$ goes to one as $n \to \infty$.

In this paper, we develop a random model for finite symmetric integral relation algebras, and prove some preliminary results.

A relation algebra is *symmetric* if $\breve{x} = x$ for all x, and *integral* if the identity $1'$ is an atom. Given a finite symmetric integral relation algebra A, atoms of A apart from $1'$ are called *diversity* atoms; if a, b, and c are diversity atoms and $a \leq b\,;c$, then abc is called a *mandatory diversity cycle*.

Definition 1. *Let $R(n, p)$ denote the probability space whose outcomes are the finite symmetric integral not-necessarily-associative relation algebras with n diversity atoms. For each diversity cycle abc, make it mandatory with probability p (and forbidden otherwise), with these choices independent of one another.*

Example 1. Let $n = 3$, and $p = \frac{1}{2}$. Given diversity atoms a, b, c, the possible diversity cycles are $aaa, bbb, ccc, abb, baa, acc, caa, bcc, cbb, abc$. The random selection of all cycles except bbb and cbb gives relation algebra 59_{65}, while the selection of only abb, acc, and bcc gives 1_{65}. Clearly, *some* selections will fail to give a relation algebra.

Theorem 1. *For any fixed $0 < p \leq 1$, the probability that $R(n, p)$ is a relation algebra goes to one as $n \to \infty$.*

Proof. We must show that $R(n, p)$ is associative, for which it suffices to show the following: for all mandatory abc and xyc, there is a z such that axz and byz are mandatory. There are $n + 2\binom{n}{2} + \binom{n}{3}$ diversity cycles, which is asymptotically $\frac{n^3}{6}$. There are thus $\binom{\frac{n^3}{6}}{2}$ possible pairs of cycles, which is asymptotically $\frac{n^6}{72}$. (This is over-counting, since some of those

---

*I thank the anonymous referee for his or her insightful comments and suggestions, which will help shape the full-length version of this article someday.

pairs won't "match up" with a common diversity atom, but it won't matter.) For any given pair abc, xyc, the probability that, for a particular atom z, axz and byz are not both mandatory is $1 - p^2$. The probability that *no* such z works is then $\Pi_z(1 - p^2)$. Hence the overall probability of failure of associativity is bounded above by

$$\sum_{\substack{abc \\ xyc}} \prod_z (1 - p^2) = \sum_{\substack{abc \\ xyc}} (1 - p^2)^n,$$

which is asymptotically $\frac{n^6}{72}(1 - p^2)^n$, which goes to zero for fixed p. (Technically, we must also show that any two diversity atoms are in a cycle together; this probability goes to one similarly.)   □

Now we turn to the question of representability. We use the fact that having a flexible atom is sufficient for representability over a countable set.

Theorem 2. *Let $p \geq n^{\overline{\binom{n+1}{2}}}$. Then the expected number of flexible atoms in $R(n, p)$ is at least one.*

Proof. Given an atom z, the probability that it is flexible is $p^{\binom{n+1}{2}}$, since all of the $\binom{n+1}{2}$ cycles involving z must be mandatory. Then by linearity of expectation we have

$$\mathbb{E}[\text{number of flexible atoms}] = \sum_z p^{\binom{n+1}{2}} = np^{\binom{n+1}{2}}.$$

Set $p \geq n^{\frac{-1}{\binom{n+1}{2}}}$. Then $np^{\binom{n+1}{2}} \geq n(n^{\frac{-1}{\binom{n+1}{2}}})^{\binom{n+1}{2}} = 1$.   □

Theorem 2 has two rather glaring shortcomings. First, it doesn't show that the probability of representability goes to one as $n \to \infty$, as one usually wants. Second, using the presence of a flexible atom as a sufficient condition for representability is overkill. It seems like it ought to be possible to strengthen Theorem 2 to prove that almost all finite symmetric integral relation algebras are representable, and a more general definition of $R(n, p)$ might allow a positive solution to problem 20 from [4]: If $RA(n)$ (respectively, $RRA(n)$) is the number of isomorphism types of relation algebras (respectively, representable relation algebras) with no more than n elements, is it the case that

$$\lim_{n \to \infty} \frac{RRA(n)}{RA(n)} = 1?$$

However, what is really desired (by this author, at least) is a notion of a *quasirandom* relation algebra. There are many graph properties, all asymptotically equivalent, that hold almost surely in $G_{n,1/2}$ and therefore can be taken as a definition of a quasirandom graph. One such example is the property of having all but $o(n)$ vertices of degree $(1 + o(1))\frac{n}{2}$. Such properties serve as proxies for "randomness".

In a similar fashion, quasirandom subsets of $\mathbb{Z}/n\mathbb{Z}$ were defined in [1]. Again, a number of properties were proved to be asymptotically equivalent. One such property is that of the characteristic function of the subset $A \subseteq \mathbb{Z}/n\mathbb{Z}$ having small (as in $o(n)$) nontrivial Fourier coefficients.

What would be a quasirandom relation algebra? Restricting attention once again to symmetric integral relation algebras, here is one possibility. For each atom a, form a graph G_a with vertices labeled with the diversity atoms, with an edge between b and c if abc is mandatory (or a loop on a if aaa is mandatory). Then call the algebra quasirandom if all but $o(n)$ of the graphs G_a are quasirandom.

Is this a good definition? I don't know. I offer it merely as an example of the sort of thing one might propose. My purpose is to start a conversation that might lead to a significant interaction between the field of relation algebra and the subfield of combinatorics that is concerned with quasirandom structures. This paper is a first step.

Here are a few problems to consider.

Problem 1. Is there a function $p(n)$ such that $R(n, p(n))$ is asymptotically the uniform distribution on symmetric integral relation algebras of order 2^{n+1}?

Problem 2. Improve the bound on p in Theorem 2.

Problem 3. Formulate several notions of quasirandomness for relation algebras, and show that they are equivalent, as in [1,2]. Maddux's work on algebras with no mandatory 3-cycles [5] suggests that the difficult part of representability lies in the 3-cycles. Results on quasirandom 3-uniform hypergraphs might be relevant.

Problem 4. First-order graph properties obey a 0-1 law in the standard uniform random graph model, i.e., every property holds with asymptotic probability 1 or asymptotic probability 0 in $G_{n,1/2}$. Does the same hold for $R(n, p)$ for some p?

The anonymous referee suggested the following alternative random model:

Definition 2. *For each finite* n, *consider a fixed finite Boolean algebra* B_n *with* n *atoms. Let* $NA(B_n)$ *be the set of finite non-associative relation algebras whose Boolean part is* B_n. *Its cardinality* $|NA(B_n)|$ *is some function of* n. *Choose an algebra* A *in* $NA(B_n)$ *uniformly at random.*

Problem 5. Are these two random models asymptotically equivalent for some $p > 0$?

Problem 6 (due to the referee). Let $p > 0$ and let Σ be a fixed finite set of equations true in all representable relation algebras. Prove that

$$\Pr[R(n, p) \models \Sigma] \to 1 \text{ as } n \to \infty.$$

References

1. F. R. K. Chung and R. L. Graham. Quasi-random subsets of Z_n. *J. Combin. Theory Ser. A*, 61(1):64–86, 1992. 10
2. F. R. K. Chung, R. L. Graham, and R. M. Wilson. Quasi-random graphs. *Combinatorica*, 9(4):345–362, 1989. 10
3. P. Erd″os and A. Rényi. On random graphs. I. *Publ. Math. Debrecen*, 6:290–297, 1959. 8
4. Robin Hirsch and Ian Hodkinson. *Relation algebras by games*, volume 147 of *Studies in Logic and the Foundations of Mathematics*. North-Holland Publishing Co., Amsterdam, 2002. With a foreword by Wilfrid Hodges. 9
5. Roger D. Maddux. Finite symmetric integral relation algebras with no 3-cycles. In *Relations and Kleene algebra in computer science*, volume 4136 of *Lecture Notes in Comput. Sci.*, pages 2–29. Springer, Berlin, 2006. 10

Multiplayer Rock-Paper-Scissors

Charlotte Aten*

Department of Mathematics, University of Rochester, NY
caten2@u.rochester.edu

1 Introduction

The game of Rock-Paper-Scissors (RPS) involves two players simultaneously choosing either rock (r), paper (p), or scissors (s). Informally, the rules of the game are that "rock beats scissors, paper beats rock, and scissors beats paper". That is, if one player selects rock and the other selects paper then the latter player wins, and so on. If two players choose the same item then the round is a tie.

A *magma* is an algebra $\mathbf{A} := (A, f)$ consisting of a set A and a single binary operation $f \colon A^2 \to A$. We will view the game of RPS as a magma. We let $A := \{r, p, s\}$ and define a binary operation $f \colon A^2 \to A$ where $f(x, y)$ is the winning item among $\{x, y\}$. This operation is given by the table below and completely describes the rules of RPS. In order to play the first player selects a member of A, say x, at the same time that the second player selects a member of A, say y. Each player who selected $f(x, y)$ is the winner. Note that it is possible for both players to win, in which case we have a tie.

$$
\begin{array}{c|ccc}
 & r & p & s \\
\hline
r & r & p & r \\
p & p & p & s \\
s & r & s & s \\
\end{array}
$$

In general we have a class of *selection games*, which are games consisting of a collection of items A, from which a fixed number of players n each choose one, resulting in a tuple $a \in A^n$, following which the round's winners are those who chose $f(a)$ for some fixed rule $f \colon A^n \to A$. We refer to an algebra $\mathbf{A} := (A, f)$ with a single basic n-ary operation $f \colon A^n \to A$ as an *n-ary magma* or an *n-magma*. We will sometimes abuse this terminology and refer to an n-ary magma $\mathbf{A}$ simply as a *magma*. Each such

---
*Thanks to Scott Kirila for pointing out the result we use in Section 2. This research was supported in part by the people of the Yosemite Valley.

game can be viewed as an n-ary magma and each n-ary magma can be viewed as a game in the same manner, providing we allow for games where we keep track of who is "player 1", who is "player 2", etc. Again note that any subset of the collection of players might win a given round, so there can be multiple player ties.

The classic RPS game has several desirable properties. Namely, RPS is, in terms we proceed to define,

1. conservative,
2. essentially polyadic,
3. strongly fair, and
4. nondegenerate.

Let $\mathbf{A} := (A, f)$ be an n-magma. We say that an operation $f\colon A^n \to A$ is *conservative* when for any $a_1, \ldots, a_n \in A$ we have that $f(a_1, \ldots, a_n) \in \{a_1, \ldots, a_n\}$[3, p.94]. Similarly we call $\mathbf{A}$ *conservative* when f is conservative. We say that an operation $f\colon A^n \to A$ is *essentially polyadic* when there exists some $g\colon \mathrm{Sb}(A) \to A$ such that for any $a_1, \ldots, a_n \in A$ we have $f(a_1, \ldots, a_n) = g(\{a_1, \ldots, a_n\})$. Similarly we call $\mathbf{A}$ *essentially polyadic* when f is essentially polyadic. We say that f is *fair* when for all $a, b \in A$ we have $|f^{-1}(a)| = |f^{-1}(b)|$. Let A_k denote the members of A^n which have k distinct components for some $k \in \mathbf{N}$. We say that f is *strongly fair* if for all $a, b \in A$ and all $k \in \mathbf{N}$ we have $|f^{-1}(a) \cap A_k| = |f^{-1}(b) \cap A_k|$. Similarly we call $\mathbf{A}$ *(strongly) fair* when f is (strongly) fair. Note that if f (respectively, $\mathbf{A}$) is strongly fair then f (respectively, $\mathbf{A}$) is fair, but the reverse implication does not hold. We say that f is *nondegenerate* when $|A| > n$. Similarly we call $\mathbf{A}$ *nondegenerate* when f is nondegenerate.

Thinking in terms of selection games we say that $\mathbf{A}$ is conservative when each round has at least one winning player. We say that $\mathbf{A}$ is essentially polyadic when a round's winning item is determined solely by which items were played, not taking into account which player played which item or how many players chose a particular item. We say that $\mathbf{A}$ is fair when each item has the same probability of being the winning item (or tying). We say that $\mathbf{A}$ is strongly fair when each item has the same chance of being the winning item when exactly k distinct items are chosen for any $k \in \mathbf{N}$. Note that this is not the same as saying that each player has the same chance of choosing the winning item (respectively, when exactly k distinct items are chosen). When $\mathbf{A}$ is *degenerate* (i.e. not nondegenerate) we have that $|A| \leq n$. In the case that $|A| \leq n$ we have that all members of $A_{|A|}$ have the same set of components. If $\mathbf{A}$ is essentially polyadic with $|A| \leq n$ it is impossible for $\mathbf{A}$ to be strongly fair unless $|A| = 1$.

Extensions of RPS which allow players to choose from more than the three eponymous items are attested historically. The French variant of RPS gives a pair of players 4 items to choose among[5, p.140]. In addition to the usual rock, paper, and scissors there is also the well (w). The well beats rock and scissors but loses to paper. The corresponding Cayley table is given below. This game is not fair, as $|f^{-1}(r)| = 3$ yet $|f^{-1}(p)| = 5$. It is nondegenerate since there are 4 items for 2 players to chose among. It is also conservative and essentially polyadic.

	r	p	s	w
r	r	p	r	w
p	p	p	s	p
s	r	s	s	w
w	w	p	w	w

There has been some recent recreational interest in RPS variants with larger numbers of items from which two players may choose. For example, the game Rock-Paper-Scissors-Spock-Lizard[1] (RPSSL) is attested in the popular culture. The Cayley table for this game is given below, with v representing Spock and l representing lizard. This game is conservative, essentially polyadic, strongly fair, and nondegenerate.

	r	p	s	v	l
r	r	p	r	v	r
p	p	p	s	p	l
s	r	s	s	v	s
v	v	p	v	v	l
l	r	l	s	l	l

It is folklore that the only "valid" RPS variants for two players use an odd number of items. Currently this is mentioned on the Wikipedia entry for Rock-Paper-Scissors without citation[6] and with a reference to a collection of such games[2]. In our language we have the following result. We give a proof of a more general statement in the next section.

Theorem 1. *Let **A** be a selection game with $n = 2$ which is essentially polyadic, strongly fair, and nondegenerate and let $m := |A|$. We have that $m \neq 1$ is odd. Conversely, for each odd $m \neq 1$ there exists such a selection game.*

In the present paper we explore selection games for more than 2 simultaneous players. We give a numerical constraint on which n-magmas of order m can be essentially polyadic, strongly fair, and nondegenerate. We use this constraint to examine permissible values of m for a fixed

n and vice versa. We describe some elementary algebraic properties of magmas which are generalized RPS games.

2   RPS Magmas

The magmas we are interested in are those corresponding to selection games which have the four desirable properties possessed by Rock-Paper-Scissors.

Definition 1 (RPS magma). *Let* $\mathbf{A} := (A, f)$ *be an n-ary magma. When* $\mathbf{A}$ *is conservative, essentially polyadic, strongly fair, and nondegenerate we say that* $\mathbf{A}$ *is an RPS magma. When* $\mathbf{A}$ *is an n-magma of order m with these properties we say that* $\mathbf{A}$ *is an RPS*(m, n) *magma. We also use RPS and RPS*(m, n) *to indicate the classes of such magmas.*

Note that in order for $\mathbf{A}$ to be fair we need that the number of items (m) divides the number of tuples $|A^n| = m^n$, which is always the case. This is certainly not a sufficient condition, as we have seen, for example, that the French variant of RPS is unfair.

Our first theorem generalizes directly to selection games with more than 2 players.

Theorem 2. *Let* $\mathbf{A}$ *be a selection game with n players and m items which is essentially polyadic, strongly fair, and nondegenerate. For all primes* $p \leq n$ *we have that* $p \nmid m$. *Conversely, for each pair* (m, n) *with* $m \neq 1$ *such that for all primes* $p \leq n$ *we have that* $p \nmid m$ *there exists such a selection game.*

Proof. Since $\mathbf{A}$ is nondegenerate we must have that $m > n$. Since $\mathbf{A}$ is strongly fair we must have that $|f^{-1}(a) \cap A_k| = |f^{-1}(b) \cap A_k|$ for all $k \in \mathbf{N}$. As the m distinct sets $f^{-1}(a) \cap A_k$ for $a \in A$ partition A_k and are all the same size we require that $m \mid |A_k|$. When $k > n$ we have that $A_k = \varnothing$ and obtain no constraint on m.

When $k \leq n$ we have that A_k is nonempty. As we take $\mathbf{A}$ to be essentially polyadic we have that $f(x) = f(y)$ for all $x, y \in A_k$ such that $\{x_1, \ldots, x_n\} = \{y_1, \ldots, y_n\}$. Let B_k denote the collection of unordered sets of k distinct elements of A. Note that the size of the collection of all members $x \in B_k$ such that $\{x_1, \ldots, x_n\} = \{z_1, \ldots, z_k\}$ for distinct $z_i \in A$ does not depend on the choice of distinct z_i. This implies that for a fixed $k \leq n$ each of the m items must be the winner among the same number of unordered sets of k distinct elements in A. We have that $|B_k| = \binom{m}{k}$ so we require that $m \mid |B_k| = \binom{m}{k}$ for all $k \leq n$.

Let

$$d(m,n) := \gcd\left(\left\{\binom{m}{k} \mid 1 \leq k \leq n\right\}\right).$$

Since $m \mid \binom{m}{k}$ for all $k \leq n$ we must have that $m \mid d(m,n)$. Joris, Oestreicher, and Steinig showed that when $m > n$ we have

$$d(m,n) = \frac{m}{\text{lcm}(\{k^{\varepsilon_k(m)} \mid 1 \leq k \leq n\})}$$

where $\varepsilon_k(m) = 1$ when $k \mid m$ and $\varepsilon_k(m) = 0$ otherwise[4, p.103]. Since we have that $m \mid d(m,n)$ and $d(m,n) \mid m$ it must be that $m = d(m,n)$ and hence

$$\text{lcm}\left(\left\{k^{\varepsilon_k(m)} \mid 1 \leq k \leq n\right\}\right) = 1.$$

This implies that $\varepsilon_k(m) = 0$ for all $2 \leq k \leq n$. That is, no k between 2 and n inclusive divides m. This is equivalent to having that no prime $p \leq n$ divides m, as desired.

It remains to show that such games **A** exist when $m \neq 1$ and for all primes $p \leq n$ we have that $p \nmid m$. By this assumption we have that $k \nmid m$ whenever $2 \leq k \leq n$. Since

$$\binom{m}{k} = \frac{m!}{(m-k)!k!} = m\frac{(m-1)\cdots(m-k+1)}{k(k-1)\cdots(2)}$$

and none of the factors of $k!$ divide m it must be that $m \mid \binom{m}{k}$ for each $2 \leq k \leq n$. This implies that $m \mid |B_k|$ for each $k \leq n$ so for each $k \leq n$ we can partition B_k into m subcollections $C_k := \{C_{k,r}\}_{r \in A}$ indexed on the m elements of A, each with $|C_{k,r}| = \frac{\binom{m}{k}}{m}$. With respect to this collection of partitions $C := \{C_k\}_{1 \leq k \leq n}$ we define an n-ary operation $f \colon A^n \to A$ by $f(a_1, \ldots, a_n) := r$ when $\{a_1, \ldots, a_n\} \in C_{k,r}$ for some $k \in \{1, \ldots, n\}$. This map is well-defined since each $\{a_1, \ldots, a_n\}$ contains exactly k distinct elements for some $k \in \{1, \ldots, n\}$ and thus belongs to a unique member of one of the partitions C_k. In order to see that the resulting magma $\mathbf{A} := (A, f)$ is essentially polyadic choose some $a_0 \in A$ and let $g \colon \text{Sb}(A) \to A$ be given by

$$g(U) := \begin{cases} r & \text{when } (\exists k)U \in C_{k,r} \\ a_0 & \text{when } |U| > n \end{cases}.$$

By construction we have that $f(a_1, \ldots, a_n) = g(\{a_1, \ldots, a_n\})$ for all a_1, $\ldots$, a_n in A. The choice of a_0 was immaterial. We now show that **A** is strongly fair. Given $r \in A$ we have that $f(a_1, \ldots, a_n) = r$ with $(a_1, \ldots, a_n)$ in A_k when $\{a_1, \ldots, a_n\} \in C_{k,r}$. Note that the number of members of A_k

whose coordinates form the set $\{a_1, \ldots, a_n\}$ is the same as the number of members of A_k whose coordinates form the set $\{b_1, \ldots, b_n\}$ for some other $(b_1, \ldots, b_n) \in A_k$. This implies that each of the $|f^{-1}(r) \cap A_k|$ have the same size for a fixed k and hence $\mathbf{A}$ is strongly fair. To see that $\mathbf{A}$ is nondegenerate observe that if $1 \neq m < n$ then there is some prime p dividing m. Since $p \leq m < n$ is prime we require that $p \nmid m$, a contradiction. We see that an essentially polyadic, strongly fair, nondegenerate n-ary magma always exists when $m \neq 1$ and for all primes $p \leq n$ we have that $p \nmid m$.

We have given a description of all possible essentially polyadic, strongly fair, nondegenerate n-magmas. There is always at least one RPS magma for every $n \geq 2$, although for brevity we refrain from demonstrating this.

3   Items as a Function of Players and Vice Versa

Our numerical condition on the existence of an $\mathrm{RPS}(m, n)$ magma allows us to analyze how many items m can be used by a fixed number n of players in such a game. For the $n = 2$ case we see that the only prime $p \leq 2$ is 2 so $2 \nmid m$. For $n = 3$ we find that $2 \nmid m$ and $3 \nmid m$. As $2 \nmid m$ we have that $m \pmod 6 \in \{1, 3, 5\}$. As $3 \nmid m$ we have that $m \pmod 6 \in \{1, 2, 4, 5\}$. Combining these conditions we see that $\mathrm{RPS}(m, 3)$ algebras can only exist for $m \equiv 1 \pmod 6$ or $m \equiv 5 \pmod 6$. Our example of an RPS 3-magma of order 5 was the smallest possible and we see that the next largest RPS 3-magmas have order $m = 6 + 1 = 7$. A similar analysis can be used to obtain a constraint on m modulo the product of all primes $p \leq n$ for any fixed n. Since 2 and 3 are also the only primes $p \leq 4$ we find that $\mathrm{RPS}(m, 4)$ magmas can only exist for $m \equiv 1 \pmod 6$ or $m \equiv 5 \pmod 6$. In the case of $n = 5$, however, we obtain that $m \pmod{30} \in \{1, 7, 11, 13, 17, 19, 23, 29\}$. The smallest RPS 5-magma thus has order 7.

Our numerical condition also allows us to fix the number of items m and ask how many players n may use that number of items. By our previous work this question is answered immediately.

Theorem 3. *Given a fixed m there exists an $\mathrm{RPS}(m, n)$ magma if and only if $n < t(m)$ where $t(m)$ is the least prime dividing m.*

Proof. Suppose that there exists such a magma. We know that in this case $p \nmid m$ for all $p \leq n$. Since $t(m)$ is a prime dividing m we must have that

$n < t(m)$. Conversely, if $n < t(m)$ then all primes $p \leq n$ are less than $t(m)$ and as such do not divide m. It is possible to show that RPS(m, n) magmas always exist when this condition is met.

Our result implies that the only RPS$(2, n)$ magma has $n = 1$. Up to isomorphism this is the magma (A, f) with $A = \{a, b\}$ and $f : A \to A$ the identity map. There are RPS$(3, n)$ magmas for $n \leq 2$. The game of RPS is one such RPS$(3, 2)$ magma. There are RPS$(4, n)$ magmas only for $n = 1$. Up to isomorphism this is again a set A with $|A| = 4$ and $f = \mathrm{id}_A$. There are RPS$(5, n)$ magmas for $n \leq 4$. The game of RPSSL is one such RPS$(5, 2)$ magma. In the full version of this paper we will use a $\mathbb{Z}_5$ action to obtain an example of a RPS$(5, 3)$ magma. Since $2 \mid 6$ the only RPS$(6, n)$ magmas are unary. We can perform a similar analysis for any fixed number of items m.

4   Algebraic Properties of RPS Magmas

We give some basic algebraic properties of RPS magmas. Note that both the original RPS magma and the magma for the French variant are contained in the magma for RPSSL so subalgebras of RPS magmas may or may not be RPS magmas. Observe that any subset of the universe of a conservative magma is a subuniverse. For example, $\{r, s, l\}$ is a subuniverse for RPSSL and the corresponding subalgebra satisfies the numerical condition necessary for RPS magmas (being binary and of order 3), yet the corresponding subalgebra fails to be strongly fair.

The class of RPS magmas is as far from being closed under products as possible.

Theorem 4. *Let **A** and **B** be nontrivial RPS n-magmas with $n > 1$. The magma $\mathbf{A} \times \mathbf{B}$ is not an RPS magma.*

Proof. We show that $\mathbf{A} \times \mathbf{B}$ cannot be conservative. Let $\mathbf{A} := (A, f)$ and let $\mathbf{B} := (B, g)$. Let $x_1, \ldots, x_n \in A$ be distinct and let $y_1, \ldots, y_n \in B$ be distinct. Since f and g are conservative we have that $f(x) = x_i$ and $g(y) = y_j$ for some i and j. It follows that $(f \times g)((x_1, y_1), \ldots, (x_n, y_n)) = (x_i, y_j)$. Either $i \neq j$, in which case (x_i, y_j) is not one of the (x_i, y_i) and $\mathbf{A} \times \mathbf{B}$ is not conservative, or $i = j$. In this latter case we have that f is essentially polyadic so $f(\sigma(x)) = f(x)$ for any permutation σ of the x_i. This implies that

$$f(x_2, \ldots, x_n, x_1) = f(x_1, \ldots, x_n) = x_i$$

so

$$(f \times g)((x_2, y_1), \ldots, (x_n, y_{n-1}), (x_1, y_n)) = (x_i, y_i).$$

Now (x_i, y_i) does not appear among the arguments of $f \times g$ so again we see that $\mathbf{A} \times \mathbf{B}$ cannot be conservative and hence cannot be an RPS magma.

References

1. Rock paper scissors spock lizard, http://www.samkass.com/theories/RPSSL.html, (web page). Date accessed: January 16, 2018 14
2. Rock paper scissors variants, http://www.umop.com/rps.htm, (web page). Date accessed: January 16, 2018 14
3. Bergman, C.: Universal Algebra: Fundamentals and Selected Topics. Chapman and Hall/CRC (2011) 13
4. Joris, H., Oestreicher, C., Steinig, J.: The greatest common divisor of certain sets of binomial coefficients. Journal of Number Theory 21(1), 101 – 119 (1985) 16
5. Umbhauer, G.: Game Theory and Exercises. Routledge Advanced Texts in Economics and Finance, Routledge, 1 edn. (2016) 14
6. Wikipedia, https://en.wikipedia.org/wiki/Rock-paper-scissors, (web page). Date accessed: January 16, 2018 14

Joins and Maltsev Products
of Congruence Permutable Varieties

Clifford Bergman

Department of Mathematics, Iowa State University, Ames, Iowa 50011
cbergman@iastate.edu

Keywords: Maltsev product, congruence-permutable, idempotent, variety, quasivariety

Abstract. Let $\mathcal{A}$ and $\mathcal{B}$ be idempotent varieties and suppose that the variety $\mathcal{A} \vee \mathcal{B}$ is congruence-permutable. Then the Maltsev product $\mathcal{A} \circ \mathcal{B}$ is also congruence-permutable.

A group **G** is called an extension of **A** by **B** if there is a normal subgroup **N** of **G** such that $\mathbf{N} \cong \mathbf{A}$ and $\mathbf{G}/\mathbf{N} \cong \mathbf{B}$. In a series of papers, Bernard and Hanna Neumann explored the properties of the class $\mathcal{A}\mathcal{B}$ of groups, each of which is an extension of a member of the class $\mathcal{A}$ by a member of the class $\mathcal{B}$. They restricted their attention to the case that both $\mathcal{A}$ and $\mathcal{B}$ are varieties of groups (and $\mathcal{A}\mathcal{B}$ is defined to consist only of groups). Among other things, they proved that $\mathcal{A}\mathcal{B}$ is again a variety, that $(\mathcal{A}\mathcal{B})\mathcal{C} = \mathcal{A}(\mathcal{B}\mathcal{C})$, and that $\mathcal{A}\mathcal{B}$ is locally finite if both $\mathcal{A}$ and $\mathcal{B}$ are locally finite. They also described the full set of equations that hold in $\mathcal{A}\mathcal{B}$ in terms of those that hold in $\mathcal{A}$ and in $\mathcal{B}$. A full accounting of their results can be found in [9, Chap. 2].

In [7] A. I. Maltsev considered this construction in a very general context. Among his observations he showed that if $\mathcal{A}$ and $\mathcal{B}$ are quasivarieties of finite similarity type then the Maltsev product, which we denote $\mathcal{A} \circ \mathcal{B}$, is again a quasivariety. Moreover, this product contains both $\mathcal{A}$ and $\mathcal{B}$. Consequently, $\mathcal{A} \vee \mathcal{B} \subseteq \mathcal{A} \circ \mathcal{B}$ (join in the lattice of quasivarieties.) We reproduce Maltsev's construction, specifically for quasivarieties, in Definition 01 below.

Unfortunately (and in contrast to the situation for groups), it is not the case that the Maltsev product of two varieties is generally closed under homomorphic images. To address this failure Maltsev introduced a further restriction by requiring his algebras to be *polarized*. A class $\mathcal{C}$ is polarized if there is a basic unary operation symbol that is constant on every member of C and that constant is an idempotent element of the algebra. This constant is called the *pole* of the algebra. Note that the pole of a group is its identity element, the pole of an algebra is unique (if it exists), and a congruence class of a polarized algebra is a subalgebra if and only if it is the congruence class of the pole. Maltsev proved that if C is a congruence-permutable polarized variety, then for any two subvarieties $\mathcal{A}$ and $\mathcal{B}$ the class $(\mathcal{A} \circ \mathcal{B}) \cap C$ is again a variety.

Recently interest in universal algebra has turned in a somewhat different direction, towards idempotent algebras. It is easy to see that the Maltsev product of two idempotent quasivarieties is again idempotent. Freese and McKenzie [4] consider the preservation of various properties under the product. While they show that a number of important Maltsev conditions are preserved, congruence-permutability is not one of them.

In this short paper we provide some context for this failure. The main result shows that for idempotent varieties $\mathcal{A}$ and $\mathcal{B}$, if $\mathcal{A} \vee \mathcal{B}$ is congruence-permutable then so is $\mathcal{A} \circ \mathcal{B}$. Combining this with Maltsev's argument described above, if $\mathcal{A} \vee \mathcal{B}$ is congruence-permutable, then $\mathcal{A} \circ \mathcal{B}$ is a variety.

1 Maltsev Products

Classes of algebras are always assumed to be of some single fixed similarity type and closed under isomorphic image. A *quasivariety* is a class closed under subalgebra, product, and ultraproduct. Equivalently, under subalgebra and reduced product. See [3, Theorem 2.25]. A quasivariety is a *variety* if it is closed under homomorphic images. For an algebra $\mathbf{A}$, $\mathrm{Sub}(\mathbf{A})$ denotes the set of subuniverses of $\mathbf{A}$. For all other unfamiliar notions of universal algebra consult [2].

Since we work sometimes in the lattice of varieties and sometimes in the lattice of quasivarieties, we shall use the notation $\mathcal{A} \vee \mathcal{B}$ for the smallest variety containing $\mathcal{A} \cup \mathcal{B}$ and $\mathcal{A} \vee_{\mathbf{Q}} \mathcal{B}$ for the smallest quasivariety. Note that if $\mathcal{A}$ is a quasivariety then its closure under homomorphic images, $\mathbf{H}(\mathcal{A})$, is the variety generated by $\mathcal{A}$.

Congruence classes play a double role in the context of Maltsev products: as elements of a quotient algebra and as (potential) subalgebras. It may be helpful to use separate notations for the congruence class of an element a modulo the congruence θ to distinguish these roles. We shall write $[a]_\theta$ when this congruence class is being treated as a subset, and continue to write a/θ for the corresponding element of the quotient algebra.

01. **Definition** Let $\mathcal{A}$ and $\mathcal{B}$ be quasivarieties. The *Maltsev product of $\mathcal{A}$ and $\mathcal{B}$* is

$$\mathcal{A} \circ \mathcal{B} = \{\, \mathbf{R} : (\exists \theta \in \mathrm{Con}(\mathbf{R}))\ \mathbf{R}/\theta \in \mathcal{B} \text{ and}$$
$$(\forall r \in R)\ [r]_\theta \in \mathrm{Sub}(\mathbf{R}) \implies [r]_\theta \in \mathcal{A} \,\}.$$

If $\mathcal{A}$ and $\mathcal{B}$ are subquasivarieties of the quasivariety $\mathcal{C}$ then we write $\mathcal{A} \circ_{\mathcal{C}} \mathcal{B} = (\mathcal{A} \circ \mathcal{B}) \cap \mathcal{C}$.

An algebra is called *idempotent* if every singleton subset is a subuniverse. A class of algebras is idempotent if every member algebra is idempotent. Observe that in an idempotent algebra every congruence class $[a]_\theta$ is a subalgebra. We summarize the basic properties of the Maltsev product in the following theorem.

Theorem 1. *Let $\mathcal{A}$ and $\mathcal{B}$ be quasivarieties.*

1. *If the similarity type is finite, or if $\mathcal{B}$ is idempotent, then $\mathcal{A} \circ \mathcal{B}$ is a quasivariety. Moreover $\mathcal{A} \vee_{\mathbf{Q}} \mathcal{B} \subseteq \mathcal{A} \circ \mathcal{B}$.*
2. *If $\mathcal{A}$ and $\mathcal{B}$ are idempotent then $\mathcal{A} \circ \mathcal{B}$ is idempotent.*
3. *If $\mathcal{A}$ and $\mathcal{B}$ are idempotent subvarieties of a congruence-permutable quasivariety $\mathcal{C}$ then $\mathcal{A} \circ_{\mathcal{C}} \mathcal{B}$ is a variety.*

In [7] Maltsev proved 1(3) under the assumption that $\mathcal{C}$ is polarized rather than idempotent. However the proof is essentially the same in the idempotent case. A proof of Theorem 1 is also provided in [1].

2   Congruence-permutability of the Maltsev Product

In [4, Example 2.1] Freese and McKenzie exhibit idempotent varieties $\mathcal{B}_0$ and $\mathcal{B}_1$, both of which are congruence-permutable, but their join, $\mathcal{B}_0 \vee \mathcal{B}_1$, fails to be congruence-permutable. It follows from Theorem 1(1) that $\mathcal{B}_0 \circ \mathcal{B}_1$ can not be congruence-permutable. As we show in Theorem 2, this is the only obstacle to permutability of the Maltsev product.

Theorem 2. *Let $\mathcal{A}$ and $\mathcal{B}$ be idempotent varieties. If $\mathcal{A} \vee \mathcal{B}$ is congruence-permutable, then so is $\mathcal{A} \circ \mathcal{B}$.*

Recall that a variety (in fact a quasivariety) is congruence-permutable if and only if there is a ternary term $q(x, y, z)$ (a Maltsev term) such that the equations $q(x, x, y) \approx q(y, x, x) \approx y$ hold. The proof of Theorem 2 hinges on the observation that if $\mathcal{A} \vee \mathcal{B}$ is congruence-permutable then there is a single term q that simultaneously acts as a Maltsev term on $\mathcal{A}$ and on $\mathcal{B}$. A key role is played by the following result [5, Lemma 2.8] of Kearnes and Tschantz.

Lemma 1. *Let W be an idempotent variety that is not congruence-permutable. If $\mathbf{F} = \mathbf{F}_W(x,y)$ is the 2-generated free algebra in W, then $\mathbf{F}$ has subuniverses U and V such that*

1. *$x \in U, y \in V$;*
2. *$y \notin U, x \notin V$, and*
3. *$S = (U \times F) \cup (F \times V)$ is a subuniverse of $\mathbf{F} \times \mathbf{F}$.*

Proof (Proof of Theorem 2). Let q be a Maltsev term for $A \vee B$. Assume $A \circ B$ is not congruence-permutable. We shall derive a contradiction. Let $W = \mathbf{H}(A \circ B)$. Since A and B are idempotent so is W. Certainly W is not congruence-permutable so we can apply Lemma 1 to W.

So set $\mathbf{F} = \mathbf{F}_W(x,y)$. Let U and V be the subuniverses provided by the lemma and $S = (U \times F) \cup (F \times V)$. Since $\mathbf{F}$ is free and $W = \mathbf{H}(A \circ B)$, we have $\mathbf{F} \in A \circ B$. Hence there is a congruence λ on $\mathbf{F}$ such that $\mathbf{G} = \mathbf{F}/\lambda \in B$, $\mathbf{X} = [x]_\lambda \in A$ and $\mathbf{Y} = [y]_\lambda \in A$. Of course $x \in X$ and $y \in Y$.

Let $a = (x,x)$, $b = (x,y)$, $c = (y,y)$, and $d = (y,x)$. Note that $a,b,c \in S$ while $d \notin S$. We shall derive a contradiction by showing that, in fact, $d \in S$.

Let $d' = q^{\mathbf{F}^2}(a,b,c) = (p_1,p_2)$. Then $a,b,c \in S$ implies $d' \in S$ as well. From the definition of S we must have either $p_1 \in U$ or $p_2 \in V$. Without loss of generality let us assume that

$$p_2 \in V. \tag{1}$$

Now from the definitions of a, b, c, and d', we have $p_1 = q^{\mathbf{F}}(x,x,y)$. But $\mathbf{G} = \mathbf{F}/\lambda \in B$ and q is a Maltsev term for B, hence,

$$p_1/\lambda = q^{\mathbf{G}}(x/\lambda, x/\lambda, y/\lambda) = y/\lambda,$$

i.e., $p_1 \lambda y$. Thus

$$p_1 \in Y. \tag{2}$$

Similarly, $p_2/\lambda = q^{\mathbf{G}}(x/\lambda, y/\lambda, y/\lambda) = x/\lambda$, so

$$p_2 \in X. \tag{3}$$

Now let $e = (x,p_2) \in U \times F \subseteq S$. Define $e' = q^{\mathbf{F}^2}(d',e,a) = (p_3,p_4)$. Then e' is a member of S. As before, $p_3/\lambda = q^{\mathbf{G}}(p_1/\lambda, x/\lambda, x/\lambda) = p_1/\lambda$, so

$$p_3 \in Y. \tag{4}$$

From (3), $p_2, x \in X$, hence $p_4 = q^{\mathbf{F}}(p_2, p_2, x) = q^{\mathbf{X}}(p_2, p_2, x) = x$ since q is a Maltsev term for $\mathbf{X} \in A$.

Finally, let $f_1 = (y, p_2)$ and $f_2 = (p_3, p_2)$. Then $f_1, f_2 \in F \times V \subseteq S$ by (1). Therefore $q^{\mathbf{F}^2}(f_1, f_2, e') \in S$. But

$$q^{\mathbf{F}^2}(f_1, f_2, e') = (q^{\mathbf{Y}}(y, p_3, p_3), q^{\mathbf{X}}(p_2, p_2, x)) = (y, x) = d$$

proving that $d \in S$. Contradiction.   □

Corollary 1. *Let $\mathcal{A}$ and $\mathcal{B}$ be idempotent varieties and suppose that $\mathcal{A} \vee \mathcal{B}$ is congruence-permutable. Then $\mathcal{A} \circ \mathcal{B}$ is variety.*

Proof. Let $\mathcal{C} = \mathcal{A} \circ \mathcal{B}$. By Theorem 1(1), $\mathcal{C}$ is a quasivariety, and by Theorem 2, it is congruence-permutable. Therefore $\mathcal{A} \circ \mathcal{B} = \mathcal{A} \circ_C \mathcal{B}$ is a variety by Theorem 1(3).   □

Corollary 2. *Let $\mathcal{A}$ be an idempotent, congruence-permutable variety. Then $\mathcal{A} \circ \mathcal{A}$ is congruence-permutable. Furthermore $\mathcal{A} \circ \mathcal{A}$ is a variety.*

Unfortunately Theorem 2 does not provide a recipe for finding a Malstev term for $\mathcal{A} \circ \mathcal{B}$ given the term for $\mathcal{A} \vee \mathcal{B}$. We have managed this in one case. Let $\mathcal{S}q$ denote the variety of *squags*. This is the variety of binars defined by the identities

$$x \cdot x \approx x, \quad x \cdot y \approx y \cdot x, \quad x \cdot (x \cdot y) \approx y.$$

This variety is obviously idempotent. It is congruence-permutable with Maltsev term $q(x, y, z) = y \cdot (x \cdot z)$. Therefore by Corollary 2, $\mathcal{S}q \circ \mathcal{S}q$ must be a congruence-permutable variety. In [6] Li showed that a Maltsev term for $\mathcal{S}q \circ \mathcal{S}q$ is $p(x, y, z) = (x(z(xy))) \cdot (z(x(zy)))$.

Problem 1. Find an equational base for $\mathcal{S}q \circ \mathcal{S}q$. Is this variety finitely based?

While we have stated Theorem 2 for varieties it could just as easily have been stated for quasivarieties.

Corollary 3. *Let $\mathcal{A}$ and $\mathcal{B}$ be idempotent quasivarieties. If $\mathcal{A} \vee_Q \mathcal{B}$ is congruence-permutable then $\mathcal{A} \circ \mathcal{B}$ is congruence-permutable.*

Proof. Suppose that $\mathcal{C} = \mathcal{A} \vee_Q \mathcal{B}$ is a congruence-permutable quasivariety. Then $\mathcal{W} = \mathbf{H}(\mathcal{C})$ is a congruence-permutable variety. But it is easy to check that $\mathcal{W} = \mathbf{H}(\mathcal{A}) \vee \mathbf{H}(\mathcal{B})$. Therefore by Theorem 2, $\mathbf{H}(\mathcal{A}) \circ \mathbf{H}(\mathcal{B})$ is congruence-permutable. Consequently $\mathcal{A} \circ \mathcal{B} \subseteq \mathbf{H}(\mathcal{A}) \circ \mathbf{H}(\mathcal{B})$ is congruence-permutable as well.   □

Lemma 1 seems to be quite important in its own right. For example we have the following very striking result. Let $\mathbf{S}_2$ denote the 2-element semilattice.

Theorem 3 (Kearnes). *Let $\mathcal{W}$ be a variety of commutative, idempotent binars, and assume that $\mathbf{S}_2 \notin \mathcal{W}$. Then $\mathcal{W}$ is congruence-permutable.*

Proof. Assume to the contrary that $\mathcal{W}$ is not congruence-permutable. Let U and V be the subuniverses of $\mathbf{F}$ promised by Lemma 1. We claim that either U or V is an ideal of $\mathbf{F}$. (U is an ideal means that $u \in U$ and $a \in F$ implies $ua, au \in U$.) Suppose not. Then (because of commutativity) there are $u \in U, v \in V, a, b \in F$ such that $ua \notin U$ and $bv \notin V$. But then

$$(u, b) \cdot (a, v) = (ua, bv) \notin (U \times F) \cup (F \times V)$$

contradicting the assertion that S is a subuniverse.

Therefore, without loss of generality, we may assume that U is an ideal of $\mathbf{F}$. Then $B = U \cup \{y\}$ is a subuniverse of $\mathbf{F}$ and $\mathbf{B}$ has a congruence θ with two blocks, namely U and $\{y\}$. Consequently $\mathbf{S}_2 \cong \mathbf{B}/\theta \in \mathcal{W}$, a contradiction. □

References

1. Clifford Bergman, *Notes on quasivarieties and Maltsev products*, http://www.math.iastate.edu/cbergman/manuscripts/maltsevprods.pdf. 22
2. Clifford Bergman, *Universal algebra. Fundamentals and selected topics*, Pure and Applied Mathematics (Boca Raton), vol. 301, CRC Press, Boca Raton, FL, 2012. MR 2839398 (2012k:08001) 21
3. S. Burris and H. P. Sankappanavar, *A course in universal algebra*, Springer-Verlag, New York, 1981, Available from http://www.math.uwaterloo.ca/~snburris/htdocs/ualg.html. 21
4. Ralph Freese and Ralph McKenzie, *Maltsev families of varieties closed under join or Maltsev product*, Algebra Universalis **77** (2017), no. 1, 29–50. MR 3602782 21, 22
5. Keith A. Kearnes and Steven T. Tschantz, *Automorphism groups of squares and of free algebras*, Internat. J. Algebra Comput. **17** (2007), no. 3, 461–505. MR 2333368 22
6. J. Li, *Congruence n-permutable varieties*, Ph.D. thesis, Iowa State University, 2017, Graduate Theses and Dissertations. 15355. 24
7. Anatoliĭ Ivanovič Malcev, *Multiplication of classes of algebraic systems*, Siberian Math. J. **8** (1967), 254–267, Translated in [8]. 20, 22
8. _____, *The metamathematics of algebraic systems. Collected papers: 1936–1967*, North-Holland Publishing Co., Amsterdam-London, 1971, Translated, edited, and provided with supplementary notes by Benjamin Franklin Wells, III, Studies in Logic and the Foundations of Mathematics, Vol. 66. MR 0349383 25
9. H. Neumann, *Varieties of groups*, Springer–Verlag, Berlin, 1967. 20

Asymptotic Properties of Finite Groupoids

David M. Clark

Department of Mathematics, State University of New York, New Paltz 12561
clarkd@newpaltz.edu

Abstract. Recent work has shown that randomizing algorithms drawn from the methods of evolutionary computation can efficiently find a term representing a given term operation on an idemprimal groupoid provided the groupoid has two other properties: NSR and AC. This work motivates a deeper study of these two properties which prove to be worthy of investigation quite apart from their application to these algorithms.

The **Term Generation Problem** (TGP) is the problem of finding a term to represent an arbitrary finitary operation on a given finite algebra in those cases where the given operation is in fact a term operation. Natural targets for the TGP are *primal algebras* (for which all operations are term operations) and *idemprimal algebras* (for which all idempotent-preserving operations are term operations). A computationally efficient method is given in [3] to decide if a finite algebra is primal by using the Universal Algebra Calculator of [6]. This method leads to a simple recursive solution to the TGP for primal algebras, but it was shown in [4] that the recursions generally produce unfeasibly large terms.

There remains no deterministic solution to the TGP that will produce human scale terms in human scale time. But the authors of [2], [4], [5] and [9] have found a variety of randomizing algorithms to do this when the algebra is a groupoid. They have shown that these algorithms, which draw on methods of evolutionary computation, are highly efficient when the given groupoid is both *idemprimal* (**IPr**) and *term continuous* (**TC**).

A finite groupoid **G** is *term continuous* if its term to term operation function is continuous relative to certain appropriate metrics. This concept was introduced in [2], where it was proven that a groupoid **G** is TC provided that it has two other previously unidentified properties: it is *asymptotically complete* (**AC**) and it has *no subgroupoid with a separating relation* (**NSR**). Consequently the above cited randomizing algorithms will efficiently solve the TGP for **G** if **G** has three properties:

IPr, NSR and AC.

This note is not about algorithms to solve the TGP itself, but rather about the properties of having NSR and being AC that have proven to be of independent interest. The significance of the facts that NSR and AC imply TC and that IPr, NSR and AC imply that our randomizing algorithms will efficiently solve the TGP depend on answers to two questions about those properties.

(Q1) How can we tell if a groupoid has the property?
(Q2) How common is the property among all finite groupoids?

If we can effectively test for these properties and we find that the outcomes are frequently positive, we will have a basis to conclude that the above cited results are more significant.

For the property IPr, we have Murskiĭ's theorem [7] saying that **almost all** finite groupoids are IPr, that is, the proportion of IPr groupoids among n-element groupoids approaches one as n goes to infinity. (See [1], Chapter 6, for an English translation.) This gives us an optimal answer to Q2. But there is no known deterministic algorithm to decide if a finite groupoid is IPr. However, as a partial answer to Q1, we now have a randomizing method that can often confirm that a groupoid is IPr. First, the groupoid can be efficiently checked to see that it has no non-trivial congruences, automorphisms or proper subalgebras. According to [8] and [10], it is then idemprimal if and only if it has a discriminator term. If it is indeed IPr and happens to have NSR and be AC, then the algorithms of [4] or [5] will produce a discriminator term – thereby confirming that it is IPr. This observation suggests that we turn to asking questions Q1 and Q2 about NSR and AC.

A non-empty, irreflexive, symmetric binary relation σ on a finite groupoid **G** is a **separating relation** if it is preserved by left and right multiplication by elements of **G**, that is,

$$(a,b) \in \sigma \longrightarrow (ac, bc) \in \sigma \text{ and } (ca, cb) \in \sigma$$

for all $a, b, c \in G$. We say **G** has **no separating relations (NSR)** if no subgroupoid of **G** has a separating relation. An efficient paper-and-pencil test is given in [2] to determine whether or not **G** has a separating relation. We let $\rho_1 := \{(a,a) \mid a \in G\}$ and

$$\rho_{n+1} := \rho_n \cup \{(a,b) \in G^2 \mid (ac, bc) \in \rho_n \text{ or } (ca, cb) \in \rho_n \text{ for some } c \in G\}.$$

It follows immediately by induction that, for all n, no element of ρ_n is in a separating relation. Since $\rho_n \subseteq \rho_{n+1}$ for all n, there is an n for which

$\rho_n = \rho_{n+1}$. If $\rho_n = G^2$, then **G** has no separating relation. Otherwise, it is easy to check that $G^2 \setminus \rho_n$ is a separating relation on **G**. Most people who have applied this algorithm to a few randomly generated finite groupoids have been led to the following unproven conjecture.

Conjecture 1. *Almost all finite groupoids have NSR.*

To define asymptotic completeness, let **G** be a finite groupoid, let $a \in G$, let k and H be positive integers, let $\mathbf{d} \in G^k$ and let X_k be the set $\{x_0, x_1, \ldots, x_{k-1}\}$ of k variables. Since the set of terms in variables X_k of height at most H is finite, we can define $\beta_{\mathbf{d},a}(H)$ be the probability that a term of height at most H in variables X_k will take value a at $\mathbf{d}$. It is shown in [2] that these probabilities can be calculated by recursion on H. As an example of these calculations, consider the groupoid **A** in Figure 1.

$$
\begin{array}{c|ccc}
* & 0 & 1 & 2 \\
\hline
0 & 2 & 1 & 2 \\
1 & 1 & 0 & 0 \\
2 & 0 & 0 & 1 \\
\end{array}
$$

Fig. 1. Finite groupoid **A**.

Taking $k = 5$ and $\mathbf{d} = (2, 0, 0, 1, 2)$, we used a spreadsheet to calculate $\beta_{\mathbf{d},a}(H)$ with $a = 0, 1, 2$ for some sample choices of H to get the following table.

H	$\beta_{\mathbf{d},0}(H)$	$\beta_{\mathbf{d},1}(H)$	$\beta_{\mathbf{d},2}(H)$
01	0.400000	0.200000	0.400000
05	0.407940	0.337173	0.254887
10	0.393967	0.339572	0.266462
20	0.397328	0.339094	0.263578
40	0.397935	0.339010	0.263055
66	0.397951	0.339007	0.263042
67	0.397950	0.339007	0.263042
68	0.397950	0.339007	0.263042

These and further calculations suggest a conjecture.

Conjecture 2. *For the groupoid **A** and the input sequence* $\mathbf{d} = (2, 0, 0, 1, 2)$, *the probability distributions* $(\beta_{\mathbf{d},0}(H), \beta_{\mathbf{d},1}(H), \beta_{\mathbf{d},2}(H))$ *converge to a value close to* $(0.397950, 0.339007, 0.263042)$.

To date we have extensive experimental evidence of similar behaviors for other finite groupoids, but proofs in only the most trivial cases. Not all such experiments appear to lead to convergence to a stable value, but those that do not appear to lead to some kind of stable oscillation.

Of course, if $a \in G$ is not in the subgroupoid $\text{sg}(\mathbf{d})$ generated by the coordinates of $\mathbf{d}$, then $\beta_{\mathbf{d},a}(H)$ will be zero for all H. In [2] a finite groupoid was defined to be **asymptotically complete (AC)** if, for all positive integers k, all $\mathbf{d} \in G^k$ and all $a \in \text{sg}(\mathbf{d})$, the sequence $\beta_{\mathbf{d},a}$ is eventually bounded away from zero. Even if we are willing to accept the conclusions suggested by spreadsheet calculations of probability distributions, we can only do tests for a finite number of values of k. This would not be a problem if we could prove the following conjecture, also supported by extensive experiments.

Conjecture 3. *Let* **G** *be a finite groupoid and let* k *and* k' *be positive integers. Assume that* $\mathbf{d} \in G^k$, *that* $\mathbf{d}' \in G^{k'}$ *and that* $\text{sg}(\mathbf{d}) = \text{sg}(\mathbf{d}')$. *If* $a \in G$, *then* $\beta_{\mathbf{d},a}$ *is eventually bounded away from zero if and only if* $\beta_{\mathbf{d}',a}$ *is eventually bounded away from zero.*

If this conjecture were true, it would mean that we could answer Q1 for AC if we could reliably test one choice of **d** for each subgroupoid of **G**. Thus a proof of Conjecture 3 might be a step toward an affirmative answer to the following question.

Problem 4. *Is AC a decidable property of finite groupoids?*

While the evidence is nowhere nearly as compelling for AC as for NSR, it is strong enough to state another conjecture.

Conjecture 5. *Almost all finite groupoids are AC.*

Conjectures 1 and 5 could have significant consequences. If they are both true, then combining them with Murskiĭ's theorem [7] would tell us that almost all finite groupoids will efficiently yield terms under the algorithms presented in [4] and [5].

We have given here but a few examples of interesting questions and conjectures about finite groupoids that arise from our work on the Term Generation Problem. For further open questions, see the final sections of [4] and [5].

References

1. C. Bergman, *Universal Algebra: Fundamentals and Selected Topics*, CRC Press (2012). 27

2. D. Clark, Evolution of algebraic terms 1: term to term operation continuity, *International Journal of Algebra and Computation*, Vol. 23, No. 5 (2013) 1175–1205. 26, 27, 28, 29

3. D. Clark, B. Davey, J.Pitkethly, D. Rifqui, Flat unars: the primal, the semi-primal and the dualisable, *Algebra Universalis* Vol 63, No. 4 (2010), 303–329. 26

4. D. Clark, M. Keijzer, L. Spector, Evolution of algebraic terms 2: deep drilling algorithm, *International Journal of Algebra and Computation*, Vol. 26, No 6 (2016) 1141–1176. 26, 27, 29

5. D. Clark, L. Spector, Evolution of algebraic terms 3: term continuity and beam algorithms, (under review). 26, 27, 29

6. R. Freese, E. Kiss, M. Valeriote, UACalc, a Universal Algebra Calculator, Available at: www.uacalc.org, (2011). 26

7. V. L. Murskiǐ, Konéčnaá baziruémost′ toźdéstv i drugié svojstva "počti vséh" konéčnyh algébr (A finite basis of identities and other properties of "almost all" finite algebras), *Problémy Kibérnétiki* **30** (1975), 43–56. 27, 29

8. A. Pixley, Functionally complete algebras generating distributive and permutable classes, *Math. Z.* **114** (1970), 361–372. 27

9. L. Spector, D. Clark, B. Barr, J. Klein and I. Lindsay, Genetic programming for finite algebras *Genetic and Evolutionary Computation Conference* (GECCO) 2008 Proceedings, Atlanta GA (July 2008), Editor-in-Chief Maarten Keijzer, Association for Computing Machinery (ACM), ISBN: 978-1-60558-130-9, pp 1291–1298. [Paper won first place in the GECCO 2008 Human Competitive competition.] ⟨www.genetic-programming.org/hc2011/combined.html⟩ 26

10. H. Werner, Eine Charakterisierung funktional vollständiger Algebren, *Arch. Math.* (Basel) **21** (1970), 381-385. 27

Minimal SH-approximation of semigroups

V. V. Dang, S. Yu. Korabel'shchikova, and B. F. Mel'nicov

[1] Department of Applied Sciences, Vietnam National University-HCMC,
Hochiminhcity University of Technology, Saigon, Vietnam
dangvvinh@hcmut.edu.vn
[2] Department of Algebra and Geometry, Northern (Arctic) M. V. Lomonosov
Federal University, Arkhangelsk, Russia, s.korabelsschikova@narfu.ru
[3] Department of Applied Mathematics and Informatics, Togliatti State
University, Togliatti, Russia, bf-melnikov@yandex.ru

Abstract. The problem of SH-approximation of a semigroup with respect to the predicate of a possible belonging of an element to a subsemigroup is considered. Several explicit conditions for SH-approximation with respect to this predicate are presented. We constructed a special semigroup acting the role of a minimal SH-approximation of semigroup for many predicates. This semigroup has neither identity nor additive identity. It contains an infinite number of idempotents, and the presence of each idempotent is mandatory.

Keywords: semigroup, approximation, mimimal approximation of semigroup

1 Introduction

The common concept of approximation of algebraic system was given in the research of Russian academician A. I. Mal'cev [6]. In this article, Mal'cev showed a connection between the finitely approximation of an algebraic system with respect to a given predicate and the problem of solvability of this predicate in the system. The notion of a finitely approximable semigroup is also mentioned. Some results about approximation of a semigroup were proposed by A. I. Mal'cev in the article.

The problem of finding the minimal approximation and SH-approximation of semigroups was proposed by professor M.M. Lesokhin. We study the approximation and SH-approximation of semigroups with respect to different predicates in semigroup theory, see [1]–[5]. For a given class of semigroups, we introduce a method to find a minimal semigroup of approximation. The presence of an identity and a zero element are required for some predicates of approximation, especially the predicate of membership of an element to a subsemigroup, which not only requires the presence of an identity element, but also an external attached identity. We create a special semigroup C^*, which acts as a minimal semigroup of approximation for this predicate. The semigroup C^* has not only no identity, but also a zero element. It also contains an infinite number of idempotents and the presence of every idempotent is mandatory.

2   Preliminaries

Definition 1. *Let Q be the set of all prime numbers. Let $G_p, p \in Q$ be the quasi-cyclic group of the type p^∞ with an identity e_p and with an operation denoted by $\oplus_p$. Put $C^* = \cup G_p, (p \in Q)$. Define in C^* multiplication as follows. $\forall a_p, a_q \in C^*$*

$$a_p * a_q = \begin{cases} a_p \oplus_p a_q, & \text{if } p = q; \\ a_{\max\{p,q\}}, & \text{if } p \neq q \text{ and } \max\{p,q\} > 3; \\ e_5, & \text{if } p \neq q \text{ and } \max\{p,q\} = 3. \end{cases}$$

Direct calculation shows that $C^* = (C^*, *)$ is a semigroup, a semilattice of groups $G_p, p \in Q$.

Let A and B be two algebraic structures as the same type, Φ be the set of all mappings between A and B, and P be a predicate defined on the set that consists of A, all subsets $\delta(A)$ of A and all images of A and $\delta(A)$ under the mappings from Φ.

Definition 2. *An algebraic structure A is said to be approximable by mappings from Φ with respect to P, if for a pair of subsets A_1, A_2 from A such that $P(A_1, A_2)$ is false, there exists $\varphi \in \Phi$ such that $P(\varphi(A_1), \varphi(A_2))$ is also false.*

Definition 3. *An algebraic structure A is said to be SH-approximable by mappings from Φ with respect to P, if every image of any substructure of A under a mapping from Φ is approximable with respect to this predicate.*

Definition 4. *An algebraic structure B is called a minimal SH-approximable structure for a class K with respect to P, if the following three conditions hold:*

 (i) *any structure $A \in K$ is SH-approximable with respect to P by mappings in B;*
 (ii) *if a structure S is SH-approximable with respect to P by mappings in B, then $S \in K$;*

(iii) if B_1 is a proper substructure of B, then there exists a structure $A \in K$ such that A is not SH-approximable by mappings in B_1 with respect to P.

In this article, the algebraic structures A is a semigroup and the structure B is the semigroup C^*; mappings between A and C^* are homomorphisms.

3 A minimal semigroup of approximation

Lemma 1. *If a semigroup A is approximable into C^* by homomorphisms with respect to the predicate of the possible belonging of an element to a monogenic semigroup, then A is approximable by homomorphisms into C^* with respect to the predicate "equality of two elements."*

Proof. For simplicity, let us denote by P the predicate of the possible belonging of an element to a monogenic semigroup. Suppose that $a, b \in A, a \neq b$ be two elements of A, and $[a], [b]$ be two monogenic subsemigroups generated by a and b respectively. There exist three cases:

1. $a \notin [b]$. The semigroup A is approximable with respect to P. There exists a homomorphism φ, such that $\varphi(a) \notin \varphi([b])$. It is clear that $\varphi(b) \in \varphi([b])$. Hence $\varphi(a) \neq \varphi(b)$. Consequently, A is approximable with respect to the predicate of equality of two elements.
2. Similarly, we obtain a proof for the second case $b \notin [a]$.
3. $[a] = [b]$. Because $a \neq b$, there are two natural numbers $k \neq 1, l \neq 1$, such that $a = b^k$ and $b = a^l$. Then $a = (a^l)^k = a^m$, hence $[a]$ is a finite cyclic subsemigroup. Since the index of a is 1, the monogenic subsemigroup $[a]$ is a group.

Because $a, b \in [a]$ and $a \neq b$, $ab^{-1} \neq e$, $ab^{-1} \notin [e]$, then there is a homomorphism φ into C^*, such that $\varphi(ab^{-1}) \notin \varphi([e])$. However,

$$\varphi(a)(\varphi(b))^{-1} \neq \varphi(e),$$

hence $\varphi(a) \neq \varphi(b)$. Thus, the semigroup A is approximable with respect to the predicate "equality of two elements." □

Lemma 2. *Let B be the commutative semigroup of three idempotents, e_1, e_2, and e_3, such that $e_1 = e_1 e_2 = e_1 e_3$ and $e_2 e_3 = e_2$. Then any commutative semigroup A of idempotents is approximable by homomorphisms into B with respect to the predicate of the possible belonging of an element to a subsemigroup.*

Proof. Let A' be a subsemigroup of the semigroup A and $e_0 \notin A'$. Let us denote by J_{e_0} the set of all elements $a \in A$ such that $ae_0 \neq e_0$. We will prove that J_{e_0} is an ideal of A. First, we show that J_{e_0} is a subsemigroup of A. For any two elements $a, b \in J_{e_0}$, by the definition of J_{e_0}, we have $ae_0 \neq e_0$ and $be_0 \neq e_0$. If $ab \notin J_{e_0}$, then $abe_0 = e_0$ and from $e_0 = abe_0$ implies $ae_0 = aabe_0 = abe_0 = e_0 \Rightarrow ae_0 = e_0$. That contradicts $a \in J_{e_0}$. Assume that $a \in A \setminus J_{e_0}$ and $b \in J_{e_0}$. It means that $ae_0 = e_0$ and $be_0 \neq e_0$. Then $e_0 = ae_0 \Rightarrow be_0 = abe_0 \Rightarrow e_0 \neq be_0 = abe_0$. Hence $abe_0 \neq e_0$, so that $ab \in J_{e_0}$. We can make a conclusion that J_{e_0} is an ideal of the semigroup A. Suppose that $A' \subseteq J_{e_0}$. Because $e_0 \notin J_{e_0}$, for a homomorphism $\varphi :$ $A \longrightarrow B$ is defined by: $\forall a \in A$,

$$\varphi(a) = \begin{cases} e_1, & \text{if } a \in J_{e_0}; \\ e_2, & \text{if } a \notin J_{e_0}. \end{cases}$$

we obtain that $\varphi(e_0) \notin \varphi(A')$.

Now assume that $A' \not\subseteq J_{e_0}$ implies $A' \cap (A \setminus J_{e_0}) = A''$ is a subsemigroup of $A \setminus J_{e_0}$. Remember that e_0 is an element of $A \setminus J_{e_0}$ and is a zero element of $A \setminus J_{e_0}$. Let us denote by A_0 the maximal subsemigroup in $A \setminus J_{e_0}$, that contains A'' and does not contain e_0. This subsemigroup exists, for example A''. We will prove that $(A \setminus J_{e_0}) \setminus A_0$ is an ideal of $A \setminus J_{e_0}$.

Assume that $e' \in A \setminus J_{e_0}$ and $e' \notin A_0$. Then $[e'; A_0] \neq A_0$. On the other hand, $A_0 \subset [e'; A_0]$. It means that $e_0 \in [e'; A_0]$, or equivalently, there is an element $e \in A_0$, satisfying the condition $e_0 = e'e$. Suppose that a is an arbitrary element of $(A \setminus J_{e_0}) \setminus A_0$ and a' is any element of $A \setminus J_{e_0}$. Assume that $aa' \notin (A \setminus J_{e_0}) \setminus A_0$. Then $aa' \in A_0$ and $a \notin A_0$, consequently, there exists an element $a'' \in A_0$, such that $e_0 = aa''$. Consider the product $aa'a''e_0$. We have $aa'a''e_0 = aa'a''(aa'') = (aa')a''$. Due to $aa' \in A_0$ and $a'' \in A_0$, we have $(aa')a'' \in A_0$. On the other hand $aa'a'' \in A \setminus J_{e_0}$ and the element e_0 is a zero element of the semigroup $A \setminus J_{e_0}$ implies $aa'a''e_0 = e_0$. Thus, $aa'a''e_0 = (aa')a'' \in A_0$ and $aa'a''e_0 = e_0 \notin A_0$. This contradiction gives us a reason for making a conclusion that $(A \setminus J_{e_0}) \setminus A_0$ is an ideal of $A \setminus J_{e_0}$. Direct calculation shows that $(A \setminus J_{e_0}) \setminus A_0$ is a simple ideal and a mapping ψ from A to B that is defined in the following way: $\forall a \in A$,

$$\psi(a) = \begin{cases} e_1, & \text{if } a \in J_{e_0}; \\ e_2, & \text{if } a \in (A \setminus J_{e_0}) \setminus A_0; \\ e_3, & \text{if } a \in A_0. \end{cases}$$

is a homomorphism. In addition, $\psi(e_0) = e_2$. Because $(A \setminus J_{e_0}) \setminus A_0) \cap A' = \emptyset$, $\psi(e_0) \notin \psi(A')$. Finally, A is approximable by homomorphism into B with respect to the predicate of the possible belonging of an element to a subsemigroup. □

Theorem 1. *Let K be the class of commutative, regular and periodic semigroups. The semigroup C^* is a minimal SH-approximable semigroup for the class K with respect to the predicate of the possible belonging of an element to a subsemigroup.*

Proof. 1. Let A be an element of K. Because A is a commutative, regular and periodic semigroup, $A = \cup A_e$ is a disjoint union of subgroups A_e, where e belongs to a set of all idempotents E of the semigroup A. Let a be an element of $A_{e_0} \subset A$ and A_1 be a subsemigroup of A. Assume that $a \notin A_1$. There exist two possible cases:

a. $A_{e_0} \cap A_1 = \varnothing$. Let us consider a mapping ψ from A to a set of all A_e, that is defined by: $\forall a \in A_{e_a}, \psi(a) = A_{e_a}$. Direct calculation shows that $\psi(A)$ is a semilattice. Let's denote this semilatice by J. We have $\psi(a) \in J, \psi(A_1) \subset J$ and $\psi(a) \notin \psi(A_1)$. By the lemma 3.2, J is approximable by homomorphisms into a subsemigroup $T = \{e_5, e_7, e_{11}\}$ and hence A is approximable into the semigroup C^*.

b. $A_{e_0} \cap A_1 = M \neq \varnothing$. Since A_1 is a subsemigroup and A_{e_0} is a periodic subgroup, M is a subgroup, consequently $e_0 \in M$ (where e_0 is an identity of the subgroup A_{e_0}). Since $a \notin M$, there exists a homomorphism φ_0 from A_{e_0} into a subgroup G_{p_0} of the semigroup C^*, such that

$$\varphi_0(a) \notin \varphi_0(M). \tag{1}$$

This homomorphism can be extended to a homomorphism ψ_0 of the whole semigroup A into the semigroup $G_{p_0} \cup \{e_q\}$, where $q > p_0$ in the following way: $\forall c \in A$,

$$\psi_0(c) = \begin{cases} e_q, & \text{if } c \in A_{e_c}, \text{ and } e_c e_0 \neq e_0; \\ \varphi_0(ce_0), & \text{if } c \in A_{e_c}, \text{ and } e_c e_0 = e_0. \end{cases}$$

We have to find a homomorphism from A into C^* such that $\varphi(a) \notin \varphi(A_1)$. Assume that, for any homomorphism τ of the semigroup A into C^* we always have $\tau(a) \in \tau(A_1)$, accordingly $\psi_0(a) \in \psi_0(A_1)$ and there is an element $b \in A_1$ such that $\psi_0(b) = \psi_0(a)$. Obviously $a \neq b$. Since $\psi_0(a) = \varphi_0(ae_0) = \varphi_0(a) \neq e_q, \psi_0(b) \neq e_q$. It follows that $e_b e_0 = e_0$, and hence $be_0 \in A_{e_0}$. On the other hand $b \in A_1$ and $e_0 \in A_1$ imply $be_0 \in A_1$ and $be_0 \in M$. Since $\psi_0(a) = \varphi_0(ae_0) = \varphi_0(a)$ and $\psi_0(a) = \psi_0(b), \varphi_0(a) = \psi_0(b) = \varphi_0(be_0) \in \varphi_0(M)$. It contradicts (1). Thus, there is a homomorphism from A into C^* that separates the image of the element a and the image of the subsemi-group A_1.

c. Suppose that A is approximable by homomorphisms from A into the semigroup C^*. We have to show that $A \in K$, that is A is a commutative, periodic and regular semigroup. Since A is approx-imable with respect to the predicate of the possible belonging of an element to a subsemigroup, then A is approximable with respect to a predicate of the possible belonging of an element to a mono-genic semigroup. By the lemma 3.1, the semigroup A is approx-imable with respect to the predicate "equality of two elements." On the contrary, we suppose that the semigroup A is not commu-tative. There exist two elements a and b of the semigroup, such that $ba \neq ab$. Since A is approximable with respect to the predicate "equality of two elements," there is a homomorphism φ, such that $\varphi(ba) \neq \varphi(ab)$ or $\varphi(b)\varphi(a) \neq \varphi(a)\varphi(b)$. The latter inequality does not hold in the semigroup C^*. Thus, the semigroup A is commuta-tive.

Assume that the semigroup A is not regular. There exists an element $a \in A$, such that $axa \neq a$ for any $x \in A$, hence $a \notin Aa^2$. Because A is approximable, there is a homomorphism φ, such that $\varphi(a) \notin \varphi(Aa^2)$. Since Aa^2 is a two-sided ideal of the semigroup A, $\varphi(Aa^2)$ is also a two-sided ideal of $\varphi(A)$. Assume that $\varphi(a) \in G_{p_0}$ for some prime number p_0. Because $\varphi(a^3) \in G_{p_0}$ and $\varphi(a^3) \in \varphi(Aa^2)$, $G_{p_0} \cap \varphi(Aa^2) = M \neq \emptyset$. Obviously, the set M is a subgroup of G_{p_0}. Since $\varphi(Aa^2)$ is an ideal of $\varphi(A)$, M is an ideal of $\varphi(A) \cap G_{p_0}$. We have $\varphi(A) \cap G_{p_0}$ is a subgroup of G_{p_0}, therefore $M = \varphi(Aa^2) \cap G_{p_0} = \varphi(A) \cap G_{p_0}$. Since $\varphi(a) \in \varphi(A) \cap G_{p_0}$, $\varphi(a) \in \varphi(Aa^2) \cap G_{p_0}$, $\varphi(a) \in \varphi(Aa^2)$ which is impossible.

Thus, the semigroup A is commutative and regular. By this reason, A is a disjoint union of subgroups $A = \cup A_e$. We show that A is periodic. On the contrary, we suppose that A is not periodic. Consequently, there exists an element $a \in A$, such that $[a]$ is an infinite subsemi-group. Then there exists an idempotent $e_0 \in E$ such that $a \in A_{e_0}$ and $[a] \subseteq A_{e_0}$. Since the semigroup A is regular, A_{e_0} is a group. Hence the element a has an inverse a^{-1}. If $a^{-1} \in [a]$, then there is a natural num-ber m such that $a^{-1} = a^m$, hence $a^{-1}a = a^{m+1}$ and $a^{m+1} = e$. It means that $[a]$ is a finite subsemigroup. We have found a contradiction. Thus $a^{-1} \notin [a]$. Because A is approximable, there exists a homomorphism φ from A into C^* such that $\varphi(a^{-1}) \notin \varphi([a])$. Since $\varphi(a^{-1})$ is an inverse of $\varphi(a)$, the set $\varphi([a])$ is a subgroup of C^*, so $\varphi(a^{-1}) \in \varphi(A)$. We have found a contradiction. Finally, the semigroup A is a commutative, periodic and regular semigroup.

2. Let C_1^* be a proper subsemigroup of C^*. There are two possible cases:

   a. The subsemigroup C_1^* does not contain all idempotents of the semigroup C^*. There exists a prime number p, such that $e_p \notin C_1^*$. Hence $G_p \not\subset C_1^*$. We select a cyclic group H with an order p and an identity 1_H. Obviously, $H \in K$. For any homomorphism φ from H into C_1^* and for any element $g \in H$, we have $1_{C_1^*} = \varphi(1_H) = \varphi(g^p) = (\varphi(g))^p$. Since all elements (except idempotents) of the subsemigroup C_1^* have orders that are different from p, for any element $g \in H$, $\varphi(g)$ is an idempotent and $\varphi(H)$ is a group that contains only one element. In this case, the semigroup H is not approximable into C_1^* with respect to the predicate of the possible belonging of an element to a subsemigroup.

   b. The subsemigroup C_1^* contains all idempotents of the semigroup C^*. There exists an element $c \in C^*$ and $c \notin C_1^*$. Assume that the order of the element c is p^k for some prime number p and natural number k. Let us consider the semigroup $A = G_p$. Obviously, $A \in K$. Let A_1 be a proper subgroup of A and $|A_1| < p^k$, let a be an element of A and $|a| = p^k$. Clearly, $a \notin A_1$ and for any homomorphism φ from A to C_1^*, we obtain $|\varphi(A)| < p^k$. It means that $\varphi(a) \in \varphi(A_1)$, thus A is not approximable by homomorphisms into C^*.

The theorem 3.3 has been completely proved.   □

Within the framework of this report, we only present the proof of the result for the predicate of the possible belonging of an element to a subsemigroup. Apart from the above result, we also have found the necessary and sufficient conditions with respect to Greens relations, such as ℓ - equivalency, D-equivalence and H-equivalence and other predicates such as the equality of elements, the belonging of an element to a monogenic subsemigroup, the belonging of an element to a maximal subgroup, the belonging of an element to a subgroup and so on. For diversity, we add two more results in this article without proof.

Theorem 2. *A semigroup A is approximable by homomorphisms from Φ with respect to Green relation ℓ-equivalency if and only if the semigroup A can be embedded into a semilattice of left simple semigroups.*

Theorem 3. *Let K_1 be the class of commutative, separative and periodic semigroups. The semigroup C^* is a minimum approximable semigroup for the class K_1 with respect to the predicate of the possible belonging of an element to a monogenic subsemigroup.*

4   Conclusion

The approximation of semigroups consists of three components. The first component is a set of algebraic structures such as groups, finite groups, semigroups, compact semigroups, fields, etc. The second component is a set of predicates and the last one is the set of functions such as homomorphisms, continuous characters, continuous bi-characters, etc. Changing one of these components give us a new research direction.

Finding a minimal semigroup of approximation appears naturally. When we study approximation by homomorphisms into B, we would like to find the smallest subsemigroup B_1 of the semigroup B such that A is approximable by homomorphisms into B_1.

References

1. Dang, V. V., S. Yu. Korabel'shchikova, B. F. Mel'nikov, *Semigroups approximation with respect to some ad hoc predicates*, Arctic envirenmontal research, vol. 17, pp. 133–140, 2017. 31
2. Dang, V. V., S. Yu. Korabel'shchikova, B. F. Mel'nikov, *On the problem of finding minimum semigroup of approximation*, Izvestiya Vysshikh Uchebnykh Zavedenhii, Povolzhskiy Region, pp. 88–98, 2015.
3. Golubov E. A., *The finite approximability of separable, naturally linearly ordered commutative semigroups*, Izv. Vuzov Matem. no. 2, pp. 23–31, 1969.
4. Ignateva I. V., *SH-approximation of semigroup by finite charaters*, Modern Algebra, vol. 1, Rostov-na-Donu, Russia, pp. 25–30, 1996.
5. Lesokhin M. M., E. A. Golubov, *The finite approximability of commutative semigroups*, Matem. Zap. Ural'skogo Univ, vol. 5, no. 3, pp. 82–90, 1966. 31
6. Mal'cev A. I., "Selected works," vol. 1, Nauka, Moscow, 1976. 31

Residuation algebras with functional duals

Wesley Fussner and Alessandra Palmigiano*†

1 Department of Mathematics, University of Denver, Denver, CO 80208
Wesley.Fussner@gmail.com
2 Faculty of Technology, Policy & Management, Delft University of Technology
Department of Pure and Applied Mathematics, University of Johannesburg
A.Palmigiano@tudelft.nl

Abstract. We employ the theory of canonical extensions to study residuation algebras whose associated relational structures are *functional*, i.e., for which the ternary relations associated to the expanded operations admit an interpretation as (possibly partial) functions. Providing a partial answer to a question of Gehrke, we demonstrate that no universal first-order sentence in the language of residuation algebras is equivalent to the functionality of the associated relational structures.

1 Introduction

In the context of a research program aimed at establishing systematic connections between the foundations of automata theory in computer science and duality theory in logic, in [3], Gehrke specializes extended Stone and Priestley dualities in the tradition of [5] so as to capture *topological algebras*[3] as dual spaces. Specifically, topological algebras based on Stone spaces are characterized as those relational Stone spaces, as in [5], in which the $(n + 1)$-ary relations dually corresponding to n-ary operations on Boolean algebras are *functional*, and an analogous result is obtained for topological algebras based on Priestley spaces. In particular, focusing the presentation on residuation algebras (see Definition 1), the

---

*The second author was supported by the Vidi grant 016.138.314 of the Netherlands Organization for Scientific Research (NWO), by the NWO Aspasia grant 015.008.054, and by a Delft Technology Fellowship awarded in 2013.

†We wish to thank Peter Jipsen for his careful reading and very useful comments on an earlier draft of this paper.

3For any algebraic similarity type τ, a topological algebra of type τ is an algebra of type τ in the category of topological spaces, i.e. it is a topological space endowed with continuous operations for each $f \in \tau$.

additional operations on distributive lattices are characterized for which the dual relations are functional (see [3, Proposition 3.16]). These results are formulated and proved without explicit reference to the theory of canonical extensions.

This note is motivated by a question raised in [3, end of Section 3.2], viz. whether the conditions of the statement of [3, Proposition 3.16] are equivalent to a first-order property of residuation algebras. To address this question, we have recast some of the notions and facts pertaining to residuation algebras in the language and theory of canonical extensions, which allows for these facts to be reformulated independently of specific duality-theoretic representations. Our contributions are as follows.

Firstly, we obtain a more modular and transparent understanding of how the validity of the inequality $a\backslash(b \vee c) \leq (a\backslash b) \vee (a\backslash c)$ forces the functionality of the dual relation.[4] In each setting (Boolean, distributive), the validity of this inequality forces the product of join-irreducible elements (which is a closed element, by the general theory of π-extensions of normal dual operators) to be either $\perp$ or finitely join prime (cf. Proposition 1). Moreover, prime closed elements of the canonical extension of a general lattice expansion are completely join-irreducible (see Lemma 2). The functionality of the dual relation is obtained as a consequence of these two facts, of which only the first depends on the validity of the inequality above.

Secondly, we provide a partial answer to the initial question. Specifically, functionality cannot be captured by any equational condition or quasiequational condition, since there is no first-order *universal* sentence in the language of residuation algebras (or even residuated lattices) that is equivalent to functionality (see Example 1).

Thirdly and finally, we articulate a version of [3, Proposition 3.16]—reformulated in a purely algebraic fashion—in which one of the equivalent conditions in the statement is made weaker, and the corresponding part of the proof is simplified and rectified (see Proposition 2).

2   Residuation algebras and their canonical extensions

Definition 1. *(cf. [3], Definition 3.14) A* residuation algebra *is a structure* $\mathbb{A} = (A, \backslash, /)$ *such that A is a bounded distributive lattice,* $\backslash$ *and* / *are binary*

---

[4]Note that $(a\backslash b) \vee (a\backslash c) \leq a\backslash(b \vee c)$ holds in every residuation algebra by the monotonicity of $\backslash$ in its second coordinate, and hence $a\backslash(b \vee c) \leq (a\backslash b) \vee (a\backslash c)$ is equivalent to $a\backslash(b \vee c) = (a\backslash b) \vee (a\backslash c)$.

operations on A such that \ *(resp.* /) *preserves finite (hence also empty) meets in its second (resp. first) coordinate, and for all* $a, b, c \in A$,

$$b \leq a \backslash c \quad \textit{iff} \quad a \leq c/b.$$

The canonical extension of $\mathbb{A}$ *as above is the algebra* $\mathbb{A}^\delta = (A^\delta, \backslash^\pi, /^\pi)$ *such that* A^δ *is the canonical extension of* A *(see [4, Definition 2.5]), and* $\backslash^\pi$ *and* $/^\pi$ *are the* π-*extensions of* \ *and* /, *respectively (see [4, Definition 4.1]).*

The residuation condition of the definition above implies that \ (resp. /) converts finite (hence empty) joins in its first (resp. second) coordinate into meets. Together with the meet-preservation properties mentioned in the definition above, this implies (see [4, Lemma 4.6]) that $\backslash^\pi$ and $/^\pi$ preserve *arbitrary* meets in their order-preserving coordinates and reverse *arbitrary* joins in their order-reversing coordinates. Since A^δ is a complete lattice, this implies that an operation $\cdot : A^\delta \times A^\delta \to A^\delta$ exists which is completely join-preserving in each coordinate and such that for all $u, v, w \in A^\delta$,

$$v \leq u \backslash^\pi w \quad \textit{iff} \quad u \cdot v \leq w \quad \textit{iff} \quad u \leq w /^\pi v.$$

Hence, $\mathbb{A}^\delta$ is a complete residuation algebra endowed with the structure of a complete lattice-ordered residuated groupoid. Moreover, $\cdot$ restricts to the elements of the meet-closure[5] of A in A^δ, denoted $K(A^\delta)$ (see [1, Lemma 10.3.1]).

Definition 2. *For any residuation algebra* $\mathbb{A}$ *as above, its associated relational dual structure* $\mathbb{A}_+^\delta := (J^\infty(A^\delta), \geq, R)$ *is based on the set* $J^\infty(A^\delta)$ *of the completely join-irreducible elements[6] of* A^δ *with the converse order inherited from* A^δ, *and endowed with the ternary relation* R *on* $J^\infty(A^\delta)$ *defined for* $x, y, z \in J^\infty(A^\delta)$ *by*

$$R(x, y, z) \quad \textit{iff} \quad x \leq y \cdot z.$$

Such an R *is* functional *if* $y \cdot z \in J^\infty(A^\delta) \cup \{\perp\}$ *for all* $y, z \in J^\infty(A^\delta)$, *in which case we also say that* $\mathbb{A}_+^\delta$ *is functional, and is* functional and defined everywhere *if* $y \cdot z \in J^\infty(A^\delta)$ *for all* $y, z \in J^\infty(A^\delta)$. *In this case, we say that* $\mathbb{A}_+^\delta$ *is* total.[7]

---

[5]The join-closure of A in A^δ is denoted $O(A^\delta)$.

[6]$x \in A^\delta$ is *completely join-irreducible* if $x = \bigvee S$ implies $x \in S$ for any $S \subseteq A^\delta$. If A is distributive, A^δ is completely distributive and hence completely join-irreducible elements are *completely join-prime*, i.e. for any $S \subseteq A^\delta$, if $x \leq \bigvee S$ then $x \leq s$ for some $s \in S$.

[7]Notice that functional relations as defined in [3, Definition 3.1] correspond to relations which are functional and defined everywhere in the present paper.

Group relation algebras, full relation algebras over a given set, and semi-linear residuated lattices give examples of residuation algebras whose dual structures are functional.

Notice that by allowing the possibility that $y \cdot z = \bot$, we are allowing the set $R^{-1}[y, z] := \{x \mid R(x, y, z)\}$ to be empty for some $y, z \in J^\infty(A^\delta)$. We emphasize that it is not uncommon that $y \cdot z = \bot$ for $y, z \in J^\infty(A^\delta)$. For instance, in any finite Boolean algebra, where $\backslash$ and $/$ coincide with the Boolean implication and $\cdot$ coincides with $\wedge$, the product of two distinct join-irreducible elements is $\bot$. Examples of algebras in which the product of join-irreducibles may be $\bot$ are also found among MV-algebras and Sugihara monoids. A residuation algebra $\mathbb{A}$ as above *has no zero-divisors* if $x \cdot y \neq \bot$ for all $x, y \in J^\infty(A^\delta)$.

The next two lemmas give a useful connection between the duality-theoretic perspective of [3] and the setting of canonical extensions. Specifically, they capture in a purely algebraic fashion one key property of prime filters of *general* lattices, namely that each prime filter induces a maximal filter/ideal pair, given by itself and its complement. This fact underlies why primeness implies join-irreducibility.

Lemma 1. *For any lattice L, if $k \in K(L^\delta)$ is finitely prime[8] and $o = \bigvee\{b \in L \mid b \not\geq k\}$, then $k \not\leq o$.*

Proof. By way of contradiction, suppose that $\bigwedge\{a \in L : k \leq a\} = k \leq o$. Then by compactness, there exist finite sets $A \subseteq \{a \in L : k \leq a\}$ and $B \subseteq \{b \in L : b \not\geq k\}$ such that

$$a' = \bigwedge A \leq \bigvee B = b'$$

Then $a' \geq k$, and $b' \not\geq k$ (for if not, then by the primeness of k we would have $b \geq k$ for some $b \in B$, a contradiction). But then $k \leq a' \leq b'$, so $k \leq b'$, a contradiction. This settles the lemma. $\square$

Lemma 2. *For any lattice L, if $k \in K(L^\delta)$ is finitely prime, then $k \in J^\infty(L^\delta)$.*

Proof. By denseness it is enough to show that if $k = \bigvee S$ for $S \subseteq K(L^\delta)$, then $k = s$ for some $s \in S$. Let $o = \bigvee\{a \in L \mid a \not\geq k\}$, and, toward a contradiction, assume that $s < k$ for all $s \in S$. The assumption that $S \subseteq K(L^\delta)$ implies that for each $s \in S$,

$$s = \bigwedge\{a \in L \mid a \geq s\},$$

---

[8]$u \in L^\delta$ is *finitely prime* if $u \neq \bot$ and for all $v, w \in L^\delta$, if $u \leq v \vee w$ then $u \leq v$ or $u \leq w$.

whence for all $s \in S$ there exists $a_s \in L$ such that $a_s \geq s$ and $a_s \not\geq k$. Hence, $a_s \leq o = \bigvee\{a \in L \mid a \not\geq k\}$ for each $s \in S$, and so $\bigvee\{a_s \mid s \in S\} \leq o$. Therefore,

$$o \geq \bigvee\{a_s \mid s \in S\} \geq \bigvee S = k,$$

which contradicts Lemma 1, proving the claim.   □

While the lemmas above hold for *general* lattices, the next proposition makes use of residuation algebras being based on distributive lattices.

Proposition 1. *For any residuation algebra* $\mathbb{A}$, *if* $\mathbb{A} \models a\backslash(b \vee c) \leq (a\backslash b) \vee (a\backslash c)$, *then the dual structure* $\mathbb{A}_+^\delta$ *is functional.*

Proof. The inequality $a\backslash(b \vee c) \leq (a\backslash b) \vee (a\backslash c)$ is Sahlqvist (see [1, Definition 3.5]), and hence canonical (see [1, Theorems 7.1 and 8.8]). That is, the assumption that $\mathbb{A} \models a\backslash(b \vee c) \leq (a\backslash b) \vee (a\backslash c)$ implies that $\mathbb{A}^\delta \models a\backslash(b \vee c) \leq (a\backslash b) \vee (a\backslash c)$. Our aim is to show that for all $x, y \in J^\infty(A^\delta)$, if $x \cdot y \neq \bot$ then $x \cdot y \in J^\infty(A^\delta)$. From $x, y \in J^\infty(A^\delta) \subseteq K(A^\delta)$, it follows that $x \cdot y \in K(A^\delta)$ (see discussion after Definition 1). Hence, by Lemma 2 it is enough to show that $x \cdot y$ is finitely prime. Suppose that $x \cdot y \leq \bigvee S$ for a finite subset $S \subseteq A^\delta$. By residuation, $y \leq x\backslash^\pi \bigvee S \leq \bigvee\{x\backslash^\pi s \mid s \in S\}$ (here we are using $\mathbb{A}^\delta \models a\backslash(b \vee c) \leq (a\backslash b) \vee (a\backslash c)$). By the primeness of y (here we are using distributivity), this implies that $y \leq x\backslash s$ for some $s \in S$, i.e., $x \cdot y \leq s$ for some $s \in S$, which concludes the proof.   □

The situation in which the dual relation is functional and defined everywhere is captured by the following corollary, which is an immediate consequence of the proposition above.

Corollary 1. *For any residuation algebra* $\mathbb{A}$, *if* $\mathbb{A}$ *has no zero-divisors and* $\mathbb{A} \models a\backslash(b \vee c) \leq (a\backslash b) \vee (a\backslash c)$, *then* $\mathbb{A}_+^\delta$ *is total (see Definition 2).*

Although the inequality $a\backslash(b \vee c) \leq (a\backslash b) \vee (a\backslash b)$ forces the functionality of $\mathbb{A}_+^\delta$, we observe that neither this nor any other equational condition may characterize functionality. Indeed, there is no first-order universal sentence in the language of residuation algebras that is equivalent to functionality, as the following example demonstrates.

Example 1. Consider the group $\mathbb{Z}_3$ and its complex algebra, i.e., the algebra $\mathbb{A} = (\mathcal{P}(\mathbb{Z}_3), \cap, \cup, \cdot, \backslash, /, \{0\})$, where for $A, B \in \mathcal{P}(\mathbb{Z})$,

$$A \cdot B = \{a + b \mid a \in A, b \in B\},$$

$$A\backslash B = \{c \mid A \cdot \{c\} \subseteq B\},$$

$$A/B = \{c \mid \{c\} \cdot B \subseteq A\}.$$

The algebra $\mathbb{A}$ is a finite residuation algebra (indeed, a residuated lattice), hence $\mathbb{A}^\delta = \mathbb{A}$. Moreover, $\{n\} \cdot \{m\} = \{n+m\}$ for all $n, m \in \mathbb{Z}_3$ implies that the ternary relation R on $J^\infty(\mathcal{P}(\mathbb{Z}_3))$ arising from $\cdot$ is functional and defined everywhere, hence $\mathbb{A}^\delta_+$ is functional, and even total. However, $\{\varnothing, \{0\}, \{1,2\}, \mathbb{Z}_3\}$ is the universe of a subalgebra of $\mathbb{A}$ in both the language of residuated lattices and residuation algebras in which the product of join-irreducible elements may be neither $\bot$ nor join-irreducible: for instance, $\{1,2\} \cdot \{1,2\} = \mathbb{Z}_3$ is not join-irreducible. Because the satisfaction of universal first-order sentences is inherited by subalgebras, this shows that no universal first-property in the language of residuated lattices (much less residuation algebras) may characterize the functionality of $\mathbb{A}^\delta_+$.

3   Characterizing functionality

The following proposition emends [3, Proposition 3.16]. Items (2) and (3) amount to equivalent reformulations of the corresponding items in the setting of canonical extensions. Item (1) is weaker than the corresponding item in [3, Proposition 3.16], and does not stipulate that the operation $\cdot$ gives rise to a functional relation defined everywhere (see Definition 2). The proof of (1)$\Rightarrow$(2) is essentially the same as the corresponding proof in [3, Proposition 3.16]; we observe that it goes through also under this relaxed assumption. The proof of (3)$\Rightarrow$(1) is simpler than the corresponding proof in [3, Proposition 3.16], and is where the emendation takes place.

Proposition 2. *The following conditions are equivalent for any residuation algebra* $\mathbb{A} = (L, \backslash, /)$:

1. *The relational structure* $\mathbb{A}^\delta_+$ *is functional (see Definition 2).*
2. $\forall a, b, c \in A, \forall x \in J^\infty(A^\delta) \ [x \leq a \Rightarrow \exists a'[a' \in A \text{ and } x \leq a' \text{ and } a \backslash (b \vee c) \leq (a' \backslash b) \vee (a' \backslash c)]].$
3. *For all* $x \in J^\infty(A^\delta)$, *the map* $x \backslash^\pi(\_) : O(A^\delta) \to O(A^\delta)$ *is* $\vee$-*preserving.*

Proof. (1)$\Rightarrow$(2): Let $a, b, c \in A$, and $x \in J^\infty(A^\delta)$ such that $x \leq a$. We need to find some $a' \in A$ such that $x \leq a'$ and $a \backslash (b \vee c) \leq (a' \backslash b) \vee (a' \backslash c)$. If $y \in J^\infty(A^\delta)$ and $y \leq a \backslash (b \vee c)$ i.e. $a \cdot y \leq b \vee c$, then $x \cdot y \leq b \vee c$. By assumption (1) and because in distributive lattices $x, y \in J^\infty(A^\delta)$ are prime, this implies that $x \cdot y \leq b$ or $x \cdot y \leq c$, both in the case in which $x \cdot y = \bot$ and in case $x \cdot y \neq \bot$. This can be equivalently rewritten as

$y \leq x\backslash^{\pi}b = \bigvee\{a\backslash b \mid a \in A \text{ and } x \leq a\}$ or $y \leq x\backslash^{\pi}c = \bigvee\{a\backslash c \mid a \in A \text{ and } x \leq a\}$. Since $y \in J^{\infty}(A)$, this implies that $y \leq a_y\backslash b$ or $y \leq a_y\backslash c$ for some $a_y \in A$ such that $x \leq a_y$, which implies that $y \leq (a_y\backslash b) \vee (a_y\backslash c)$. Hence, given that $a_y \in A$ and $x \leq a_y$ for all such a_y,

$$a\backslash(b \vee c) = \bigvee\{y \in J^{\infty}(A) \mid y \leq a\backslash(b \vee c)\}$$
$$\leq \bigvee\{(a\backslash b) \vee (a\backslash c) \mid a \in A \text{ and } x \leq a\}.$$

Hence, by compactness, and the antitonicity of $\backslash$ in the first coordinate,

$$a\backslash(b \vee c) \leq \bigvee\{(a_i\backslash b) \vee (a_i\backslash c) \mid 1 \leq i \leq n\} \leq (a'\backslash b) \vee (a'\backslash c)$$

where $a' := \bigwedge_{i=1}^{n} a_i \in A$ and $x \leq a'$, as required.

(2)$\Rightarrow$(3): Let $x \in J^{\infty}(A^{\delta})$ and $o_1, o_2 \in O(A^{\delta})$. We need to prove that

$$x\backslash^{\pi}(o_1 \vee o_2) \leq (x\backslash^{\pi}o_2) \vee (x\backslash^{\pi}o_2). \tag{1}$$

By definition of $\backslash^{\pi}$,

$$x\backslash^{\pi}(o_1 \vee o_2) = \bigvee\{a\backslash d \mid a, d \in A \text{ and } x \leq a \text{ and } d \leq o_1 \vee o_2\}$$

$$x\backslash^{\pi}o_1 = \bigvee\{a'\backslash b \mid a', b \in A \text{ and } x \leq a' \text{ and } b \leq o_1\}$$

$$x\backslash^{\pi}o_2 = \bigvee\{a'\backslash c \mid a', c \in A \text{ and } x \leq a' \text{ and } c \leq o_2\}$$

Thus, to prove (1) it is enough to show that, for all $a, d \in A$ such that $x \leq a$ and $d \leq o_1 \vee o_2$, some $a', b, c \subset A$ exist such that $x \leq a'$, $b \leq o_1$, $c \leq o_2$ and $a\backslash d \leq (a'\backslash b) \vee (a'\backslash c)$. From $d \leq o_1 \vee o_2 = \bigvee\{b \in A \mid b \leq o_1\} \vee \bigvee\{c \in A \mid c \leq o_2\}$ we get by compactness that $d \leq b \vee c$ for some $b, c \in A$ such that $b \leq o_1$ and $c \leq o_2$. Then, by assumption (2), $a\backslash d \leq a\backslash(b \vee c) \leq (a'\backslash b) \vee (a'\backslash c)$ for some $a' \in A$ such that $x \leq a'$, as required.

(3)$\Rightarrow$(1): Let $x, y \in J^{\infty}(A^{\delta})$. Then $x \cdot y \in K(A^{\delta})$ because of general facts about canonical extensions of maps. Hence, by Lemma 2, it is enough to show that, for all $u, v \in A^{\delta}$, if $x \cdot y \neq \bot$ and $x \cdot y \leq u \vee v$ then $x \cdot y \leq u$ or $x \cdot y \leq v$. By denseness, it is enough to prove the claim for $u, v \in O(A^{\delta})$, and by compactness, it is enough to prove the claim for $u = b \in A$ and $v = c \in A$. The assumption $x \cdot y \leq b \vee c$ can be equivalently rewritten as $y \leq x\backslash^{\pi}(b \vee c) = (x\backslash^{\pi}b) \vee (x\backslash^{\pi}c)$, the equality due to assumption (3). The primeness of y yields $y \leq x\backslash^{\pi}b$ or $y \leq x\backslash^{\pi}c$, i.e. $x \cdot y \leq b$ or $x \cdot y \leq c$, as required. □

4   Conclusion

The class of residuation algebras with functional duals is not a universal class (much less a variety) according to Example 1, but it remains open whether the property of having a functional dual may be expressed by a first-order condition in the language of residuation algebras. We pose three other questions that are implicated by the foregoing analysis. First, what is the variety generated by the class of residuation algebras with functional duals, and (in particular) do the residuation algebras with functional duals generate the variety of all residuation algebras? Second, can the treatment given in this note be extended to residuated algebraic structures with non-distributive lattice reducts? Third, given that the canonicity of Sahlqvist inequalities is key to this result, and given that the core inequality expresses the additivity of a right residual map in its order-preserving coordinates, can we extend this result to signatures of additive or multiplicative connectives on the basis of the (constructive) canonicity theory for normal and regular connectives developed in [2]? We do not presently know the answer to these questions, but their resolution would deepen our understanding of functionality and promise interesting applications.

References

1. W. CONRADIE AND A. PALMIGIANO, *Algorithmic correspondence and canonicity for non-distributive logics*, arXiv preprint arXiv:1603.08515, (2016). 41, 43
2. ———, *Constructive canonicity of inductive inequalities*, arXiv preprint arXiv:1603.08341, (2016). 46
3. M. GEHRKE, *Stone duality, topological algebra, and recognition*, Journal of Pure and Applied Algebra, 220 (2016), pp. 2711–2747. 39, 40, 41, 42, 44
4. M. GEHRKE AND J. HARDING, *Bounded lattice expansions*, Journal of Algebra, 238 (2001), pp. 345–371. 41
5. R. GOLDBLATT, *Varieties of complex algebras*, Annals of Pure and Applied Logic, 44 (1989), pp. 173–242. 39

Solving Equations – kith and kin

Paweł M. Idziak and Jacek Krzaczkowski

Jagiellonian University, Faculty of Mathematics and Computer Science
Department of Theoretical Computer Science*
idziak@tcs.uj.edu.pl, jacek.krzaczkowski@uj.edu.pl

The talk is intended to present latest achievements in searching structural algebraic conditions a finite algebra **A** has to satisfy in order to have a polynomial time algorithm that decides if an equation $s(x_1, \ldots, x_n) = t(x_1, \ldots, x_n)$, where s, t are polynomials over **A**, has a solution in **A**. We will denote this computational problem by POLSAT(A).

Note that solving equations (or systems of equations) is one of the oldest and well known mathematical problems which for centuries was the driving force of research in algebra. Let us only mention Galois theory, Gaussian elimination or Diophantine Equations. In fact, for **A** being the ring of integers this is the famous 10th Hilbert Problem on Diophantine Equations, which has been shown to be undecidable [15]. In finite realms such problems are obviously decidable in nondeterministic polynomial time.

The decision version of solving systems of equations (denoted by SYSPOLSAT) is strictly connected with Constraint Satisfaction Problem for relational structures. In [16] it has been observed that SYSPOLSAT has the same expressive power as CSP, i.e.:

- for every finite relational structure $\mathbb{D}$ there is a finite algebra $\mathbf{A}[\mathbb{D}]$ such that the problem $CSP(\mathbb{D})$ is polynomially equivalent to $SYSPOLSAT(\mathbf{A}[\mathbb{D}])$,
- for every finite algebra **A** there exists a relational structure $\mathbb{D}[\mathbf{A}]$ such that the problems $SYSPOLSAT(\mathbf{A})$ and $CSP(\mathbb{D}[\mathbf{A}])$ are polynomially equivalent.

Due to this bisimulation of SYSPOLSAT and CSP and the recent dichotomy results for CSP [2,18] one can translate the beautiful splitting conditions into the language of SYSPOLSAT. Unfortunately such a bisimulation between POLSAT and CSP is not known. All we do know is that

- for every finite relational structure $\mathbb{D}$ there is a finite algebra $\mathbf{A}[\mathbb{D}]$ such that the problem $CSP(\mathbb{D})$ is polynomially equivalent to $POLSAT(\mathbf{A}[\mathbb{D}])$.

*The project is partially supported by Polish NCN Grant # 2014/14/A/ST6/00138.

Therefore the search for a characterization of finite algebras with tractable POLSAT seems to be challenging. Nevertheless there are numerous results related to problems connected with solving equations or systems of equations over specific finite algebras. Most of them concerns well known algebraic structures as groups [3,5,8,9] rings [7,4] or lattices [17] but there are also some more general results [1,16].

A study of equation satisfiability for groups has shown [6,9] that for the group S_3 the problem has a polynomial time solution in the pure group language, while it is NP-complete after endowing S_3 with a couple of its polynomial operations. A similar phenomena occurs for the group A_4 which is also tractable in pure group language while, after adding binary commutator operation becomes NP-complete [11].

Note that restricting the language of an algebra may result in an artificial exponential inflation of the size of input (i.e. length of an equation). Indeed the short terms $t_n(x_1, x_2, \ldots, x_n) = [\ldots [[x_1, x_2], x_3] \ldots x_n]$, expressed in the pure group language of $(\cdot, {}^{-1})$ have an exponential size in n, as the number of occurrences of variables doubles whenever we pass from n to $n + 1$. Such kind of phenomena do not occur in SYSPOLSAT, as every polynomial equation $s(\overline{x}) = t(\overline{x})$ can be replaced by a system of equations of the form $y = f(x_1, \ldots, x_k)$, where f is one of the basic operations. For example for the above term t_n one can use the following representation

$$t_2 = x_1^{-1} x_2^{-1} x_1 x_2$$
$$t_3 = t_2^{-1} x_3^{-1} t_2 x_3$$
$$\vdots$$
$$t_n = t_{n-1}^{-1} x_n^{-1} t_{n-1} x_n,$$

in which $t_2, \ldots, t_n$ are treated as variables.

In case of POLSAT the sensitivity to the choice of basic operations leads to a fundamental question how to measure the size of the input for algorithms. Such a measure should not depend on how the clone of operations is presented. This independence is crucial for structural conditions that characterize algebras with fast POLSAT algorithms. Originally the size of a term/polynomial is its length which obviously corresponds to the size of its presentation by a syntax tree. This tree can be treated as a circuit in which each gate is used as an input to at most one other gate. However a term can be presented by a circuit with much more compact representation than trees. For example the terms (and therefore the corresponding syntax trees) $t_n(x_1, x_2, \ldots, x_n) = [\ldots [[x_1, x_2], x_3] \ldots x_n]$, when presented in pure group language have exponential size, while they can be presented by circuits with $6n - 5$ vertices as can be seen from Figure 1.

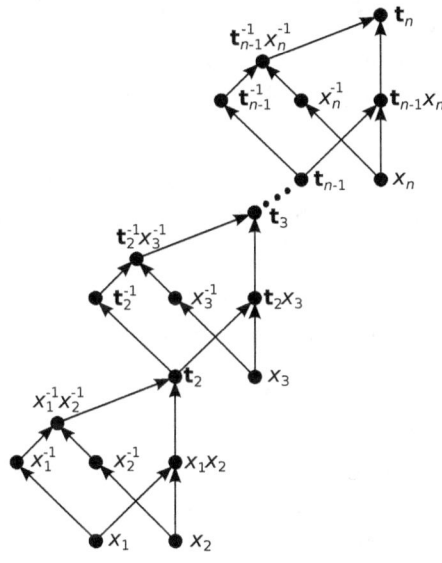

Fig. 1.

The approach by circuits proved itself to be very fruitful in our paper [13]. This paper contains an almost complete characterization of finite algebras $\mathbf{A}$ (of finite type) from congruence modular varieties with polynomial time algorithm for the following problem:

$\text{CSAT}(\mathbf{A})$
given a circuit over $\mathbf{A}$ with two output gates g_1, g_2 is there a valuation of input gates $\overline{x} = (x_1, \ldots, x_n)$ that gives the same output on g_1, g_2, i.e. $g_1(\overline{x}) = g_2(\overline{x})$.

First note that representing a polynomial by its corresponding circuit and looking at the size of this circuit (instead of the syntactic length of the polynomial) allows us to harmlessly expand the original language of the algebra $\mathbf{A}$ by finitely many of its polynomials. In fact in our intractability proofs we usually expand the language of the original algebra $\mathbf{A}$ by finitely many polynomials of $\mathbf{A}$. This allows us to code NP-complete problems more smoothly. Note that the possibility of such expansions shows that the required characterizations for CSAT can be done up to polynomial equivalence of algebras.

Actually the problem CSAT for groups had been already considered in [10] where the following characterization was obtained.

Example 1. Let **A** be a finite group. Then CSAT(**A**) is in P, whenever **A** is nilpotent [5] and NP-complete otherwise [5,10].

On the other hand the characterization of finite groups with tractable POLSAT is still far from being done. However from the existing results for POLSAT the following dichotomies can be inferred for two other classical algebraic structures.

Example 2. Let **A** be a finite ring. Then CSAT(**A**) is in P, whenever **A** is nilpotent [7] and NP-complete otherwise [4].

Example 3. Let **A** be a finite lattice. Then CSAT(**A**) $\in$ P if **A** is distributive and NP-complete otherwise [17].

Encouraged by the structural conditions in the above examples we started a systematical study of CSAT in universal algebraic setting. Some results of this study, first presented in [13], are stated in the following theorem.

Theorem 1. *Let **A** be a finite algebra from a congruence modular variety.*

1. *If **A** has no quotient **A**′ with CSAT(**A**′) being NP-complete then **A** is isomorphic to a direct product **N** × **D**, where **N** is a nilpotent algebra and **D** is a subdirect product of 2-element algebras each of which is polynomially equivalent to the 2-element lattice.*
2. *If **A** decomposes into a direct product **N** × **D**, where **N** is a supernilpotent algebra and **D** is a subdirect product of 2-element algebras each of which is polynomially equivalent to the 2-element lattice, then for every quotient **A**′ of **A** the problem CSAT(**A**′) is solvable in polynomial time.*

The nice structure obtained in (1) shows that the nilpotency and distributivity (mentioned in Examples 1-3) form a paradigm of algebras with tractable CSAT. This structure is enforced by several couples of interpretations of NP-complete problems. These interpretations make heavy use of tame congruence theory and modular commutator theory.

On the other hand the tractability proofs for both factors mentioned in (2) rely on showing that if an equation has a solution then it must have one among relatively small set S of tuples (although there may exist some other solutions outside the set S). It seems that all known polynomial time algorithms for CSAT fall into this scheme.

As we have observed in [13] our proofs of tractability for both factors in (2) show that set S depends only on the number of variables occurring in the equation but not on the structure/syntax of the equation. Now, to decide the existence of solution the algorithms evaluate the polynomials on tuples from this small set S. It seems however that in the nilpotent but not supernilpotent setting there is no chance for a polynomial time algorithm for CSAT based on this kind of ideas. In fact our results contained in [12] confirm this claim. The tools developed there allow us to show the following theorem.

Theorem 2. *If $P \neq NP$ then for some nilpotent (but not supernilpotent) algebras $\mathbf{A}$ there is no polynomial time algorithm which solves $\text{CSAT}(\mathbf{A})$ by reducing to the search space S depending only on the number of variables.*

Nevertheless both [12] and [14] contain examples of nilpotent but not supernilpotent algebras with tractable CSAT. These algorithms require a deep understanding of structure of circuits and syntax of the corresponding terms.

One of the features used to prove Theorem 2 has been based on extending satisfiability problems to algebras with infinitely many operations. However leaving the safe realm, in which only finitely many basic operations are allowed, results in several fundamental problems. To start with, note that presenting an equation $\mathbf{s}(\overline{x}) = \mathbf{t}(\overline{x})$ we need to identify the basic operations that occur in $\mathbf{s}$ or $\mathbf{t}$. In fact, even the decidability of such redefined CSAT is not clear. One way to overcome this is to have an algebra $\mathbf{A} = (A; \mathbf{f}_0, \mathbf{f}_1, \dots)$ encoded by a Turing machine $TM_{\mathbf{A}}$ which given the (k-ary) operation $\mathbf{f}_m$ and $a_1, \dots, a_k \in A$ returns $TM_{\mathbf{A}}(m, a_1, \dots, a_k) = \mathbf{f}_m(a_1, \dots, a_k)$. Such approach puts extended CSAT$(\mathbf{A})$ into NP whenever $TM_{\mathbf{A}}$ works in polynomial time. But it can be applied only to algebras with recursively enumerable set of fundamental operations.

The other way is to present an instance $\mathbf{s}(\overline{x}) = \mathbf{t}(\overline{x})$ of the problem together with the descriptions of all fundamental operations that occur in $\mathbf{s}$ or $\mathbf{t}$. Again such description may be done twofold:

- by presenting the tables of the occurring basic operations, or
- by (polynomial time) algorithms $TM_{\mathbf{f}}$ which compute the values $\mathbf{f}(a_1, \dots, a_k)$.

It may seem that presenting operations by tables is more natural as in many cases the complexity of such extended CSAT coincides with the complexity of its original (finite) version whenever the clone of an algebra is finitely generated. On the other hand presenting the tables can again be treated as an artificial inflation of the input size. Indeed, in [12] we have examples of nilpotent algebras with polynomial time CSAT when operations are presented by tables and NP-complete CSAT with TM_f-style presentation. In fact Theorem 2 relies on such examples.

One gap in the characterization provided by Theorem 1 requires a better understanding of nilpotent but not supernilpotent algebras. This includes nilpotent algebras in infinite language. The other one is our assumption about homomorphic images. This one seems even more difficult to be filled. One reason for this difficulty is that nice structure conditions expected in required characterizations are usually preserved when passing to quotient algebras. The other reason is provided by the following fact [13].

Proposition 1. *There is a finite algebra* **A** *with a congruence* α *such that* CSAT$(\mathbf{A})$ *is in* P *while* CSAT$(\mathbf{A}/\mathrm{ff})$ *is* NP-*complete.*

The algebras used to justify Fact 1 do not generate congruence modular varieties. Therefore there still is a chance that the following question has a positive answer.

Problem. Is it true that NP-completeness of CSAT for some quotient of a finite algebra **A** from a congruence modular variety implies NP-completeness of CSAT for **A** itself.

References

1. E. Aichinger and N. Mudrinski, Some applications of higher commutators in Malcev algebras, *Algebra Universalis*, **63**(2010), 367–403. 48
2. A. Bulatov, A dichotomy theorem for nonuniform CSPs, *Proceedings of the 58th Annual IEEE Symposium on Foundations of Computer Science*, 2017, 319-330. 47
3. S. Burris and J. Lawrence, Results on the equivalence problem for finite groups, *Algebra Universalis*, **52**(2005), 495–500. 48
4. S. Burris and J. Lawrence, The equivalence problem for finite rings, *Journal of Symbolic Computation*, **15**(1993) 67–71. 48, 50
5. M. Goldmann and A. Russell, The complexity of solving equations over finite groups, Proceedings of the Fourteenth Annual IEEE Conference on Computational Complexity, 1999, pp. 80–86. 48, 50
6. T. Gorazd and J. Krzaczkowski, Term equation satisfiability over finite algebras, *International Journal of Algebra and Computation*, 20(2010),1001–1020. 48

7. G. Horváth, The complexity of the equivalence and equation solvability problems over nilpotent rings and groups, *Algebra Universalis*, **66**(2011), 391-403. 48, 50

8. G. Horváth, J. Lawrence, L. Mérai and Cs. Szabó, The complexity of the equivalence problem for nonsolvable groups, *Bulletin of the London Mathematical Society*, **39**(2007), 433–438. 48

9. G. Horváth and Cs. Szabó, The Complexity of Checking Identities over Finite Groups, *International Journal of Algebra and Computation*, 16(2006), 931–940. 48

10. G. Horváth and Cs. Szabó, The extended equivalence and equation solvability problems for groups, *Discrete Mathematics & Theoretical Computer Science*, **13**(2011), 23–32. 50

11. G. Horváth and Cs. Szabó, Equivalence and equation solvability problems for the alternating group A_4, *Journal of Pure and Applied Algebra*, 216(2012), 2170–2176. 48

12. P. Idziak, P. Kawałek, J. Krzaczkowski, Expressive power, satisfiability and equivalence of circuits over nilpotent algebras, *manuscript*. 51, 52

13. P. Idziak, J. Krzaczkowski, Satisfiability in multi-valued circuits *manusript*. 49, 50, 51, 52

14. M. Kompatscher, *private communication*. 51

15. Yu. V. Matiyasevich, Enumerable Sets are Diophantine, *Soviet Mathematics Doklady*, **11**(1970),354–357 47

16. B. Larose and L. Zádori, Taylor terms, constraint satisfaction and the complexity of polynomial equations over finite algebras, *International Journal of Algebra and Computation*, **16**(2006), 563–581. 47, 48

17. B. Schwarz, The complexity of satisfiability problems over finite lattices, *Annual Symposium on Theoretical Aspects of Computer Science*, Springer 2004, 31–43. 48, 50

18. D. Zhuk, The Proof of CSP Dichotomy Conjecture, *Proceedings of the 58th Annual IEEE Symposium on Foundations of Computer Science*, 2017, 331–342. 47

Nonassociative right hoops

Peter Jipsen and Michael Kinyon*

[1] School of Computational Sciences, Chapman University, Orange, CA 92866
jipsen@chapman.edu
[2] Department of Mathematics, University of Denver, Denver, CO 80208
mkinyon@du.edu

Abstract. The class of nonassociative right hoops, or narhoops for short, is defined as a subclass of right-residuated magmas, and is shown to be a variety. These algebras generalize both right quasigroups and right hoops, and we characterize the subvarieties in which the operation $x \wedge\!\!\!\wedge y = (x/y)y$ is associative and/or commutative. Narhoops with a left unit are proved to be integral if and only if $\wedge$ is commutative, and their congruences are determined by the equivalence class of the left unit. We also prove that the four identities defining narhoops are independent.

Extended Abstract

A *residuated magma* is a partially ordered algebra $(A, \leq, \cdot, /, \backslash)$ such that $(A, \leq)$ is a poset, $\cdot$ is a binary operation and $/, \backslash$ are the right and left residuals of $\cdot$, which means the residuation property

$$x \cdot y \leq z \quad \Longleftrightarrow \quad x \leq z/y \quad \Longleftrightarrow \quad y \leq x \backslash z$$

holds for all $x, y, z \in A$. As usual, we abbreviate $x \cdot y$ by xy and adopt the convention that $\cdot$ binds stronger than $/, \backslash$. If the operation $\backslash$ is omitted then the algebra $(A, \leq, \cdot, /)$ is called a *right-residuated magma*.

Define the term $x \wedge y = (x/y)y$ and consider the following two varieties:

- A *right quasigroup* is an algebra $(A, \cdot, /)$ satisfying the identities $x \wedge y = x = (xy)/y$. Right quasigroups are precisely those right-residuated magmas for which the partial order $\leq$ is the equality relation.

*Partially supported by Simons Foundation Collaboration Grant 359872.

- A *right hoop* is an algebra $(A, \cdot, /)$ satisfying the identities $x \wedge y = y \wedge x$, $(x/x)y = y$, and $x/(yz) = (x/z)/y$. Then it turns out that x/x is a constant (denoted by 1), the operation $\cdot$ is associative, the operation $\wedge$ is a semilattice operation, 1 is the top element with respect to the semilattice order $\leq$, and $/$ is the right residual of $\cdot$ with respect to $\leq$.

Right hoops were introduced by Bosbach [1,2] under the name "left complementary semigroups" and Büchi and Owens [3] studied the case where $\cdot$ is commutative, referring to these structures as "hoops". Note that the partial order is definable in both cases, which motivates the following definition. A *nonassociative right hoop* $(A, \leq, \cdot, /)$, or *narhoop* for short, is a right-residuated magma such that for all $x, y \in A$

(N) $x \leq y \iff x \wedge y = x = y \wedge x$.

In any right-residuated magma $(x/y)y \leq x$ or equivalently $x \wedge y \leq x$ holds for all x, y, hence in a narhoop (N) implies that the identity $(x \wedge y) \wedge x = x \wedge y$ holds. This provides an alternative definition for narhoops: they are right-residuated magmas that satisfy the identity (N1) $(x \wedge y) \wedge x = x \wedge y$ and the bi-implication

(N') $x \leq y \iff x = y \wedge x$

since in the presence of (N1), if $x = y \wedge x$ then multiplying by y on the right we have $x \wedge y = (y \wedge x) \wedge y = y \wedge x = x$.

A *nonassociative left hoop* or *nalhoop* $(A, \leq, \cdot, \backslash)$ is defined dually and a *nonassociative hoop* or *nahoop* is both a narhoop and a nalhoop. Here we consider only narhoops and save the two-sided case for future research.

The two motivating varieties fit into this framework as follows.

- A narhoop $(A, \cdot, /)$ is a right quasigroup if and only if $\leq$ is the equality relation.
- A narhoop $(A, \cdot, /)$ is a right hoop if and only if the quasiequation $x \wedge y - x \Rightarrow x \leq y$ holds.

The main result of this section is that narhoops form a finitely based variety of algebras. To reduce the need for parentheses, we assume that x/y binds stronger than $x \wedge y = (x/y)y$.

Theorem 1. *Let $(A, \leq, \cdot, /)$ be a narhoop. Then the following identities hold:*

(N1) $(x \wedge y) \wedge x = x \wedge y$
(N2) $xy/y \wedge x = x$
(N3) $xz \wedge (x \wedge y)z = (x \wedge y)z$

(N4) $(x/z) \wedge (x \wedge y)/z = (x \wedge y)/z.$

Conversely, let $(A, \cdot, /)$ be an algebra with two binary operations satisfying (N1)–(N4), and define $x \leq y \iff x = y \wedge x$. Then the identies

(N5) $x \wedge xy/y = x$
(N6) $(x \wedge y)/y = x/y$
(N7) $(x \wedge y) \wedge y = x \wedge y$

hold and $(A, \leq, \cdot, /)$ is a narhoop.

Proof. Assume $(A, \leq, \cdot, /)$ is a narhoop. As noted above, the identity (N1) holds in narhoops. Right-residuated magmas also satisfy $x \leq xy/y$, hence (N2) follows from (N′).

Having a right residual implies that right-multiplication is order preserving, so $(x \wedge y)z \leq xz$ holds in all narhoops, which produces (N3). Similarly the right residual is order preserving in the first argument, hence $(x \wedge y)/z \leq x/z$ holds, and now (N4) follows from (N′).

For the converse, suppose $(A, \cdot, /)$ satisfies (N1)–(N4), and $\leq$ is defined by (N′). From (N2), (N1) and (N2) again, we get (N5):

$$x \wedge (xy/y) = (xy/y \wedge x) \wedge (xy/y) = (xy/y) \wedge x = x.$$

For (N6), replace x in (N5) by x/y to get $x/y \wedge (x \wedge y)/y = x/y$ and then use (N4). To prove (N7) multiply (N6) on the right by y.

Now reflexivity of $\leq$ follows from (N5) and (N1): $x \wedge x = (x \wedge xy/y) \wedge x = x \wedge (xy/y) = x$.

For antisymmetry, if $x \leq y$ and $y \leq x$, then $x \wedge y = x = y \wedge x$ and $y = x \wedge y$, hence $x = y$.

Transitivity is a bit more work. Suppose $x \leq y$ and $y \leq z$ so that $x \wedge y = x = y \wedge x$ and $y \wedge z = y = z \wedge y$. First, note that

$$z/x \wedge y/x = z/x \wedge (z \wedge y)/x = (z \wedge y)/x = y/x$$

using (N4) in the second equality. Now we compute

$$
\begin{aligned}
z \wedge x &= (z \wedge x) \wedge x && \text{by (N6)} \\
&= (z \wedge x) \wedge (y \wedge x) = (z/x)x \wedge (y/x)x \\
&= (z/x)x \wedge (z/x \wedge y/x)x && \text{since } z/x \wedge y/x = y/x \\
&= (z/x \wedge y/x)x && \text{by (N3)} \\
&= (z/x \wedge (z \wedge y)/x)x = ((z \wedge y)/x)x && \text{by (N4)} \\
&= (z \wedge y) \wedge x = y \wedge x = x.
\end{aligned}
$$

From $x = z \wedge x$ we deduce $x \wedge z = (z \wedge x) \wedge z = z \wedge x$ by (N1), hence $x \leq z$.

Finally, we prove $/$ is the right residual of $\cdot$ with respect to $\leq$. To do this, we verify $(x/y)y \leq x \leq xy/y$ and that $x \leq y$ implies $xz \leq yz$ and $x/z \leq y/z$ since the right residuation property is equivalent to these (quasi)identities. Note that (N2) and (N') show $x \leq xy/y$. If $x \leq y$, then (N3) gives

$$yz \wedge xz = yz \wedge (y \wedge x)z = (y \wedge x)z = xz,$$

and so (N') implies $xz \leq yz$. By the same argument, (N4) gives $x/z \leq y/z$.

To prove $(x/y)y \leq x$, or equivalently $x \wedge y \leq x$, substitute x/x for x, x for y, and $(x \wedge y)/x$ for z in (N3) to get

$$(x/x)x \wedge (x/x \wedge (x \wedge y)/x)x = (x/x \wedge (x \wedge y)/x)x.$$

Using (N4) this simplifies to $(x \wedge x) \wedge ((x \wedge y)/x)x = ((x \wedge y)/x)x$, so by (N1), (N') and reflexivity we have $x \wedge y \leq x \wedge x = x$.   □

The equational basis (N1)–(N4) for narhoops is independent as can be seen from algebras $A_i = \{0,1\}$ ($i = 1,2,3,4$) that each satisfy the axioms except for (Ni).

- In A_1, $\cdot$ is ordinary multiplication and $x/y = y$.
- In A_2, $x \cdot y = x$ and $x/y = 1$.
- In A_3, $x \cdot y$ is addition modulo 2 and $x/y = 0$ except that $1/0 = 1$.
- In A_4, $x \cdot y$ is the max operation and x/y is addition modulo 2.

In general, neither $\cdot$ nor the term operation $\wedge$ of a narhoop is associative. However $\wedge$ is associative both in right quasigroups and in right hoops. In right quasigroups, this follows from the identity $x \wedge y = x$. In right hoops, $\wedge$ turns out be a semilattice operation ([4], Lem. 4). In both cases the reduct $(A, \wedge)$ is a *left normal band*, that is, an idempotent semigroup satisfying the identity $x \wedge y \wedge z = x \wedge z \wedge y$.

If $(A, \cdot, /)$ is a narhoop and $B \subseteq A$ is closed under $\wedge$, then B inherits the order $\leq$ from A. We state the next two results in the slightly more general context of such subsets because we will need them in Theorem 4.

Theorem 2. *Let $(A, \cdot, /)$ be a narhoop and let $B \subseteq A$ be closed under $\wedge$. The following are equivalent.*

1. *$(B, \wedge)$ is a right normal band;*
2. *$(B, \wedge)$ is a semigroup;*

3. *For all* $x, y \in B$, $x \wedge (y \wedge x) = x \wedge y$

As noted, $\wedge$-reducts of right hoops are semilattices. A natural nonassociative generalization of right hoops is the variety of narhoops described in the next result. The description of the $\wedge$-reduct generalizes ([4], Lem. 4).

Theorem 3. *Let* $(A, \cdot, /)$ *be a narhoop and let* $B \subseteq A$ *be closed under* $\wedge$. *The following are equivalent.*

1. $(B, \wedge)$ *is commutative;*
2. *For all* $x, y \in B$, $x \wedge (y \wedge x) = y \wedge x$.

When these equivalent conditions hold, $(B, \wedge)$ *is a semilattice.*

In a left normal band, the identity $x \wedge y \wedge z = x \wedge z \wedge y$ essentially expresses the fact that every downset $(a] = \{x \in A \mid x \leq a\} = \{a \wedge x \mid x \in A\}$ is a subsemilattice. The same role is played by $(x \wedge y) \wedge z = (x \wedge z) \wedge y$ in narhoops.

Theorem 4. *Let* $(A, \cdot, /)$ *be a narhoop and fix* $a \in A$. *Then the downset* $(a]$ *is closed under* $\wedge$ *and is a semilattice.*

We now consider narhoops which have a left identity element.

Lemma 1. *Let* $(A, \leq, \cdot, /)$ *be a right-residuated magma such that* $x \leq y \iff x = y \wedge x$ *holds for all* $x, y \in A$. *Then*

1. x/x *is a maximal element for all* $x \in A$,
2. *the identity* $(x/x)y/y = x/x$ *holds in* A, *and*
3. *if* A *has a top element then the term* x/x *is this top element.*

Lemma 2. *Let* $(A, \cdot, /)$ *be a narhoop. The following are equivalent.*

1. $x/x = y/y$ *for all* $x, y \in A$;
2. $(x/x)y = y$ *for all* $x, y \in A$;
3. *There exists* $e \in A$ *such that* $ey = y$ *for all* $y \in A$.

When these conditions hold, the element $1 = x/x$ *is the maximum left identity element in* $(A, \leq)$.

A narhoop $(A, \cdot, /)$ is *unital* if the equivalent conditions of Lemma 2 hold. In this case as in the lemma, we denote by $1 = x/x$ the distinguished left identity element. Note that the lemma does not claim that 1 is the unique left identity element, even when $\wedge$ is commutative.

Theorem 5. *If A is finite and unital, then 1 is the unique left identity element.*

Note that in a unital narhoop A the partial order $\leq$ can be characterized in terms of 1 and $/$:

$$x \leq y \quad \Longleftrightarrow \quad x/y = 1$$

for all $x, y \in A$. Furthermore, the left identity 1 can also be used to characterize the commutativity of $\wedge$.

Theorem 6. *Let $(A, \cdot, /, 1)$ be a unital narhoop. Then:*

1. *The left unit 1 is the top element of $(A, \leq)$ if and only if $\wedge$ is commutative.*
2. *The downset $(1]$ is a subnarhoop and $((1], \wedge)$ is a semilattice.*

A congruence θ on a narhoop $(A, \cdot, /)$ is said to be *unital* if the factor
narhoop A/θ is unital. In other words, θ is unital if and only if $x/x \, \theta \, y/y$
for all $x, y \in A$. If A itself is unital then every congruence on A is unital.
For a unital congruence on an arbitrary narhoop, set

$$N_\theta = \{x \in A \mid x \, \theta \, y/y \text{ for some } y \in A\}$$
$$= \{x \in A \mid x \, \theta \, y/y \text{ for all } y \in A\},$$

where the second equality follows since θ is unital. Analogous to the re-
lationship between congruences and normal subgroups in group theory,
we now show that θ is determined by the congruence class N_θ. In order to state our characterizations concisely, we introduce six families of
mappings on a narhoop $(A, \cdot, /)$. For each $x, y \in A$, $i = 1, \ldots, 6$, define
$\phi_{i,x,y} : A \to A$ by

$$\phi_{1,x,y}(z) = (zx \cdot y)/xy, \qquad \phi_{2,x,y}(z) = (zx/y)/(x/y),$$
$$\phi_{3,x,y}(z) = (x \cdot zy)/xy, \qquad \phi_{4,x,y}(z) = (x/zy)/(x/y),$$
$$\phi_{5,x,y}(z) = xy/(x \cdot zy), \qquad \phi_{6,x,y}(z) = (x/y)/(x/zy).$$

Again keeping analogies with group theory in mind, let $\text{Inn}(A)$ denote
the transformation semigroup on A generated by these six families of
mappings.

Theorem 7. *Let θ be a unital congruence on a narhoop $(A, \cdot, /)$. Then:*

1. *N_θ is a subnarhoop of A;*
2. *For all $x, y \in A$, if $x \leq y$ and $x \in N_\theta$, then $y \in N_\theta$;*
3. *N_θ is invariant under $\text{Inn}(A)$.*

Let $(A, \cdot, /)$ be a narhoop. A nonempty subset N of A is said to be a *normal subnarhoop* of A, denoted $N \trianglelefteq A$, if the following hold:

1. N is a subnarhoop of A;
2. For all $x, y \in A$, if $x \leq y$ and $x \in N$, then $y \in N$;
3. N is invariant under $\mathrm{Inn}(A)$.

Theorem 8. *Let $(A, \cdot, /)$ be a narhoop and assume $N \trianglelefteq A$ is nonempty. Define θ_N on A by $x\,\theta_N\,y$ if and only if $x/y, y/x \in N$. Then θ_N is a unital congruence and $N_{\theta_N} = N$.*

Acknowledgments

This research was supported by the automated theorem prover PROVER9 and the finite model builder MACE4, both created by McCune [5]. We would like to thank Bob Veroff for hosting the 2016 Workshop on Automated Deduction and its Applications to Mathematics (ADAM) which is where our collaboration began.

References

1. B. Bosbach, Komplementäre Halbgruppen. Axiomatik und Arithmetik. *Fund. Math.* **64** (1969), 257–287. 55
2. B. Bosbach, Komplementäre Halbgruppen. Kongruenzen und Quotienten. *Fund. Math.* **69** (1970), 1–14. 55
3. J. R. Büchi and T. M. Owens, Complemented monoids and hoops, unpublished. 55
4. P. Jipsen, On generalized hoops, homomorphic images of residuated lattices, and (G)BL-algebras, *Soft Computing* **21**(1) (2017), 17–27. 57, 58
5. W. McCune, *Prover9 and Mace4*, version 2009-11A. http://www.cs.unm.edu/~mccune/prover9. 60

A variety $\mathcal{V}$ is congruence modular if and only if $\mathcal{V}$ satisfies $\Theta(R \circ R) \subseteq (\Theta R)^h$, for some h

Paolo Lipparini*

Dipartimento di Matematica
Viale dell'Eccellenza Scientifica Università di Roma
lipparin@axp.mat.uniroma2.it

Abstract. We present a characterization of congruence modularity by means of an identity involving a tolerance Θ and a reflexive and admissible relation R.

Keywords: Congruence modular variety; congruence identity; tolerance; reflexive and admissible relation; directed Gumm terms

More than forty years ago Nation [16] proved a result which is still intriguing today: there are non-equivalent lattice identities which nevertheless are equivalent as congruence identities in varieties. Many results of this kind followed, for example, Freese and Jónsson [4] proved that modularity is equivalent to the Arguesian identity for congruence lattices in varieties. As another example of a slightly different nature, it is an almost immediate consequence of the arguments in the proof of Lampe's Lemma [5] and of the construction of an affine-modulo-abelian term in Taylor [17] that every m-permutable variety satisfies a non-trivial congruence lattice identity. See [13] for a short history of the result and for another proof. See also Kearnes and Nation [12]. More results about congruence identities and further references can be found in Day and Freese [3], Freese and McKenzie [6], Gumm [7], Hobby and McKenzie [8], Jónsson [9] and Kearnes and Kiss [11].

It turns out that frequently tolerances, and sometimes just reflexive and admissible relations, are at work behind the scene even when results about congruences are considered. This is particularly evident, for example, in Czédli, Horváth and Lipparini [2], Jónsson [9, p. 370] or Tschantz [18]. Here we present a proof that a variety $\mathcal{V}$ is congruence modular if and only if there is some natural number h such that $\Theta(R \circ R) \subseteq (\Theta R)^h$

*Work performed under the auspices of G.N.S.A.G.A. Work partially supported by PRIN 2012 "Logica, Modelli e Insiemi." We thank the students of Tor Vergata University for stimulating discussions.

holds in every algebra in $\mathcal{V}$, for every tolerance Θ and every reflexive and admissible relation R. The proof given here is slightly simpler (but less general) than the proof presented in [14].

Our notation is as follows. Juxtaposition denotes intersection. If T is a binary relation, we let T^h denote the relational composition $T \circ T \circ T \circ \cdots \circ T$ with h factors, that is, with $h - 1$ occurrences of $\circ$. Moreover, T^* denotes the transitive closure of T.

Theorem 1. *For every variety $\mathcal{V}$, the following conditions are equivalent.*

1. *$\mathcal{V}$ is congruence modular.*
2. *There is some natural number h such that $\Theta(R \circ R) \subseteq (\Theta R)^h$ holds in every algebra in $\mathcal{V}$, for every tolerance Θ and every reflexive and admissible relation R.*
3. *The inclusion $\Theta R^* \subseteq (\Theta R)^*$ holds in every algebra in $\mathcal{V}$, for every tolerance Θ and every reflexive and admissible relation R.*
4. *The identity $\alpha \Phi^* = (\alpha \Phi)^*$ holds in every algebra in $\mathcal{V}$, for every congruence α and every tolerance Φ.*

Proof. The proof that (1) implies (2) relies heavily on a recent result by Kazda, Kozik, McKenzie and Moore [10]. There the authors showed that a variety $\mathcal{V}$ is congruence modular if and only if, for some k, $\mathcal{V}$ has $k + 1$ *directed Gumm terms*, that is, terms $p, j_1, \ldots, j_k$ satisfying the following set of equations.

$$x = p(x, z, z) \tag{DG1}$$
$$p(x, x, z) = j_1(x, x, z) \tag{DG2}$$
$$x = j_i(x, y, x), \qquad \text{for } 1 \leq i \leq k, \tag{DG3}$$
$$j_i(x, z, z) = j_{i+1}(x, x, z) \qquad \text{for } 1 \leq i < k \tag{DG4}$$
$$j_k(x, y, z) = z \tag{DG5}$$

Notice that we have given the definition of directed Gumm terms in the reversed order, in comparison with [10]. Now suppose that $\mathbf{A}$ is an algebra in $\mathcal{V}$, Θ is a tolerance of $\mathbf{A}$ and R is a reflexive and admissible relation on $\mathbf{A}$. If $a, c \in A$ and $(a, c) \in \Theta(R \circ R)$, then $a \Theta c$ and $a R b R c$, for some $b \in A$. By [10] and (1) we have terms satisfying (DG1)-(DG5) above. Let us compute

$$a = p(a, p(aab), p(aab)) \; R \; p(a, p(abb), p(aac)) = p(a, a, p(aac)),$$
$$a = p(a, a, a) = p(a, a, p(aaa)) \; \Theta \; p(a, a, p(aac)),$$

where elements in bold are those moved by R or Θ and we have used (DG1). Moreover, $p(a, a, p(aac)) = j_1(a, a, j_1(aac))$, by (DG2), hence

$$a \; \Theta R \; j_1(a, a, j_1(aac)). \tag{5}$$

Next, for $\ell = 1, \ldots, k - 1$, we have

$$j_\ell(a, \boldsymbol{a}, c) \; R \; j_\ell(a, \boldsymbol{b}, c) \; R \; j_\ell(a, \boldsymbol{c}, c) =^{\text{(DG4)}} j_{\ell+1}(a, a, c),$$
$$j_\ell(a, a, c) = j_\ell(j_\ell(a\boldsymbol{a}c), b, j_\ell(\boldsymbol{a}ac)) \; \Theta \; j_\ell(j_\ell(a\boldsymbol{a}a), b, j_\ell(\boldsymbol{c}ac)) = j_\ell(a, b, c)$$
$$j_\ell(a, b, c) = j_\ell(j_\ell(abc), c, j_\ell(a\boldsymbol{b}c)) \; \Theta \; j_\ell(j_\ell(a\boldsymbol{b}a), c, j_\ell(c\boldsymbol{b}c)) = j_\ell(a, c, c),$$

where in the last two equations we have repeatedly used (DG3). Compare Czédli and Horváth [1]. Hence

$$j_\ell(a, a, c) \; \Theta R \; j_\ell(a, b, c) \; \Theta R \; j_\ell(a, c, c) = j_{\ell+1}(a, a, c).$$

Concatenating, we get $j_1(a, a, c) \; (\Theta R)^{2(k-1)} \; j_k(a, a, c) =^{\text{(DG5)}} c$. Then

$$j_1(a, a, j_1(aac)) \; (\Theta R)^{2(k-1)} \; j_1(a, a, c) \; (\Theta R)^{2(k-1)} \; c.$$

Finally, using (5), we have $a \; (\Theta R)^{1+4(k-1)} \; c$, hence (2) holds with $h = 1 + 4(k - 1) = 4k - 3$. Let us remark that a slightly better value for h is provided in [14].

(2) $\Rightarrow$ (3) First we shall show that if

$$\Theta(R \circ R) \subseteq (\Theta R)^h \tag{6}$$

holds in some algebra $\mathbf{A}$, for every reflexive and admissible relation R, then, for every $n \geq 1$, also

$$\Theta R^{2^n} \subseteq (\Theta R)^{h^n} \tag{7}$$

holds in $\mathbf{A}$, for every reflexive and admissible relation R. This is proved by induction on n. The basis $n = 1$ is the assumption (6). If (7) holds for some n, then

$$\Theta R^{2^{n+1}} = \Theta(R^{2^n} \circ R^{2^n}) \subseteq^{(6)} (\Theta R^{2^n})^h \subseteq^{(7)} ((\Theta R)^{h^n})^h = (\Theta R)^{h^{n+1}},$$

where we have used the fact that if R is reflexive and admissible, then R^i is reflexive and admissible, for every $i \geq 1$.

Now we can prove (3). If $(a, c) \in \Theta R^*$, then $(a, c) \in \Theta R^i$, for some i, hence $(a, c) \in \Theta R^{2^n}$, for some sufficiently large n, since R is reflexive. Then by (7), $(a, c) \in (\Theta R)^{h^n} \subseteq (\Theta R)^*$. Again, notice that, given $k + 1$

directed Gumm terms, [14] provides a much better bound for ΘR^{2^n}, in comparison with the bound $(\Theta R)^{(4k-3)^n}$ given by the present proof.

$(3) \Rightarrow (4)$ The inclusion $\alpha R^* \supseteq (\alpha R)^*$ is trivial, since α is a congruence, hence transitive. The reverse inclusion follows trivially from (3).

$(4) \Rightarrow (1)$ Let α, β, γ be congruences and let Φ be the tolerance $\alpha\gamma \circ \beta \circ \alpha\gamma$. Then $\Phi^* = \beta + \alpha\gamma$, where $+$ denotes join in the lattice of congruences. Moreover, $\alpha\Phi = \alpha\gamma \circ \alpha\beta \circ \alpha\gamma$. Indeed, if $a \; \alpha \; d$ and $a \; \Phi \; d$, then there are elements b and c such that $a \; \alpha\gamma \; b \; \beta \; c \; \alpha\gamma \; d$. Since $b \; \alpha \; a \; \alpha \; d \; \alpha \; c$, we get $b \; \alpha \; c$, hence $(a, d) \in \alpha\gamma \circ \alpha\beta \circ \alpha\gamma$. From $\alpha\Phi = \alpha\gamma \circ \alpha\beta \circ \alpha\gamma$ we obtain $(\alpha\Phi)^* = \alpha\beta + \alpha\gamma$, hence from (4) and $\Phi^* = \beta + \alpha\gamma$ we get $\alpha(\beta + \alpha\gamma) = \alpha\beta + \alpha\gamma$. □

We do not know whether we still get a condition equivalent to congruence modularity if in Condition (2) in Theorem 1 we consider a reflexive and admissible relation S in place of a tolerance Θ. In any case, the varieties for which Condition (2) in Theorem 1 holds with S in place of Θ satisfy a much cleaner identity.

Proposition 1. *For every variety $\mathcal{V}$, the following conditions are equivalent.*

1. *$S(R \circ R) \subseteq (SR)^h$, for some h;*
2. *$S(R \circ R) \subseteq (SR)^*$;*
3. *$S^*R^* = (SR)^*$,*

where each condition is intended to hold in every algebra in $\mathcal{V}$ for all reflexive and admissible relations S and R.

Proof. $(1) \Rightarrow (2)$ is trivial.

$(2) \Rightarrow (1)$ is standard, though not completely usual. Suppose that (2) holds in the free algebra generated by 3 elements x, y and z and let S, R, respectively, be the smallest reflexive and admissible relations containing (x, z), respectively, (x, y) and (y, z). Then $(x, z) \in S(R \circ R)$, hence, by (2), $(x, z) \in (SR)^*$, thus $(x, z) \in (SR)^h$, for some h. This is witnessed by appropriate terms, which also witness that $S(R \circ R) \subseteq (SR)^h$ holds throughout $\mathcal{V}$. See, e. g., [15] for many similar arguments.

$(3) \Rightarrow (2)$ is trivial.

If (2) holds, then an argument similar to the proof of $(2) \Rightarrow (3)$ in Theorem 1 shows

$$SR^* \subseteq (SR)^*, \tag{8}$$

for all R and S. Applying (8) twice, we have

$$S^*R^* \subseteq^{(8)} (S^*R)^* = (RS^*)^* \subseteq^{(8)} (RS)^{**} = (SR)^*.$$

The inclusion $S^*R^* \supseteq (SR)^*$ is trivial. □

References

1. CH G. Czédli, E. K. Horváth, *Congruence distributivity and modularity permit tolerances*, Acta Univ. Palack. Olomuc. Fac. Rerum Natur. Math. **41** (2002), 39–42. 63

2. G. Czédli, E. Horváth, P. Lipparini, *Optimal Mal'tsev conditions for congruence modular varieties*, Algebra Universalis **53** (2005), 267–279. 61

3. A. Day, R. Freese, *A characterization of identities implying congruence modularity, I*, Canad. J. Math. **32** (1980), 1140–1167. 61

4. R. Freese, B. Jónsson, *Congruence modularity implies the Arguesian identity*, Algebra Universalis **6** (1976), 225–228. 61

5. R. Freese, W. A. Lampe, W. Taylor, *Congruence lattices of algebras of fixed similarity type. I*, Pacific J. Math **82** (1979), 59–68. 61

6. R. Freese, R. McKenzie, *Commutator theory for congruence modular varieties*, London Mathematical Society Lecture Note Series, vol. 125, Cambridge University Press, Cambridge, 1987. Second edition available online at http://math.hawaii.edu/~ralph/Commutator/. 61

7. H.-P. Gumm, *Geometrical methods in congruence modular algebras*, Mem. Amer. Math. Soc. **45**, 1983. 61

8. D. Hobby, R. McKenzie, *The structure of finite algebras*, Contemporary Mathematics, vol. 76, American Mathematical Society, Providence, RI, 1988. 61

9. B. Jónsson, *Congruence varieties*, Algebra Universalis 10 (1980), 355–394. 61

10. A. Kazda, M. Kozik, R. McKenzie, M. Moore, *Absorption and directed Jónsson terms*, arXiv:1502.01072 (2015), 1–17. 62

11. K. A. Kearnes, E. W. Kiss, *The shape of congruence lattices*, Mem. Amer. Math. Soc. **222**, 2013. 61

12. K. A. Kearnes, J. B. Nation, *Axiomatizable and nonaxiomatizable congruence prevarieties*, Algebra Universalis **59** (2008), 323–335. 61

13. P. Lipparini, *Every m permutable variety satisfies the congruence identity $\alpha\beta_h = \alpha\gamma_h$*, Proc. Amer. Math. Soc. **136** (2008), 1137–1144. 61

14. P. Lipparini, *Relation identities equivalent to congruence modularity*, arXiv:1704.05274, 1–9 (2017). 62, 63, 64

15. P. Lipparini, *Unions of admissible relations*, arXiv:1704.02476v3 (2018), 1–16. 64

16. J. B. Nation, *Varieties whose congruences satisfy certain lattice identities*, Algebra Universalis **4** (1974), 78–88. 61

17. W. Taylor, *Some applications of the Term Condition*, Algebra Universalis **14** (1982), 11-24. 61

18. S. T. Tschantz, *More conditions equivalent to congruence modularity*, in *Universal algebra and lattice theory (Charleston, S.C., 1984)*, 270–282, Lecture Notes in Math., **1149**, Springer, Berlin, 1985. 61

The undecidability of lattices of equational theories of large signature

George F. McNulty

Department of Mathematics, University of South Carolina, Columbia, SC
mcnulty@math.sc.edu

1 Introduction

The elementary theory of the lattice $\mathcal{L}_\Delta$ of all equational theories of signature Δ was shown to be undecidable in [4], provided the signature Δ is large in the sense that it provides at least one operation symbol of rank at least two or else provides at least two operation symbols of rank one. Indeed, they show that the elementary theory is hereditarily undecidable.

It is our object, in this study, to bring this result more sharply into focus. At the center of the reasoning of Burris and Sankappanavar is the fact, established by [3], that the lattice of all equivalence relations on any given finite set can be embedded as an interval in $\mathcal{L}_\Delta$, for all large signatures Δ. By an old result of Sachs [19], this means that the equational theory of such $\mathcal{L}_\Delta$ is the same as the equational theory of the variety of all lattices. This equational theory, the equational theory of lattices, was shown to be decidable by Thoralf Skolem in 1920 [20,21], a result later rediscovered by [23]. Thus, the *equational theory* of $\mathcal{L}_\Delta$ is decidable, but its *elementary theory* is undecidable.

In an effort to find the point along the spectrum from the equational theory to the elementary theory where undecidability enters, we offer here three further proofs of the undecidability of $\mathcal{L}_\Delta$. They rely on the substantial body of work concerning elementary definability in $\mathcal{L}_\Delta$ accomplished by Jaroslav Ježek in the decade following the theorem of Burris and Sankappanvar.

In particular, we are interested in what might be called:

Hilbert's Tenth Problem for $\mathcal{L}_\Delta$

Is there an algorithm that, upon input of a finite set of equations in the language of lattice theory, will determine whether the set of equations has a solution in $\mathcal{L}_\Delta$?

Of course, Hilbert did not pose this problem. Rather he posed the problem in which the ring $\langle \mathbb{Z}, +, \cdot, -, 0, 1 \rangle$ of integers replaces the lattice $\mathcal{L}_\Delta$.

A negative answer to this problem would amount to an assertion apparently stronger than the undecidability of the $\exists^*$-theory of $\mathcal{L}_\Delta$. We do not even know if that theory is undecidable. We can, however, prove that the $\forall^*\exists^*\forall^*$-theory of $\mathcal{L}_\Delta$ is hereditarily undecidable.

2   The approach via recursive inseparability

Roughly speaking, this method starts with a class $\mathcal{K}$ of structures which exhibits sufficiently strong undecidability properties and constructs an effective method for defining, with parameters, each structure in $\mathcal{K}$ in some model of the target theory T. The result is that T is hereditarily undecidable.

This method was pioneered by Alfred Tarski around 1940, see [22]. It was substantially widened in scope in the mid-1960's by [5] and put into a polished form by [18]. In that form it is sometimes referred to as the Rabin-Scott method. A detailed exposition of this method can be found in chapters 15 and 16 of Monk's text [14]. An enhanced version of this method is laid out in the work [2]. However, we only need a very simple version.

Let Σ be a set of elementary sentences and $\mathcal{K}$ be a class of structures of the same language as Σ. We say that $\mathcal{K}$ is *recursively inseparable relative to* Σ provided there is no recursive set S such that all the validities which belong to Σ also belong to S but that no sentence of Σ which fails in some structure in $\mathcal{K}$ can belong to S. That is, the set of validities in Σ cannot be recursively separated from the set of sentences of Σ which are rejected in $\mathcal{K}$. Notice that, in this event, if Γ is any set of sentences true in $\mathcal{K}$ such that all the validities in Σ also belong to Γ, then Γ cannot be a recursive set. So the theory of $\mathcal{K}$ exhibits a strong form of hereditary undecidability. We specialize this a bit. By a Σ-theory we mean a set T of sentences, each belonging to Σ, so that any sentence in Σ that is a logical consequence of T must already belong to T. We will say that a Σ-theory T is Σ-*hereditarily undecidable* provided each Σ-subtheory of T is undecidable. For the most part, we will take Σ to be the set of all sentences of some particular syntactic form—for example all the existential-universal sentences of the language. In this case we refer to $\exists^*\forall^*$-theories and to $\exists^*\forall^*$-hereditary undecidability.

James Schmerl has proven that, in the language with one binary relation symbol $\leq$, the class of all finite lattice-ordered sets is recursively inseparable relative to the set of all $\exists^*\forall^*$-sentences. This result can be found in Appendix A of Manuel Lerman's 1983 monograph [10]. The

1975 work of Burris and Sankappanavar used similar methods for the language of lattices where $\mathcal{K}$ is the class of all finite partition lattices and Σ is the set of all sentences in the language. We draw an easy consequence of Schmerl's theorem.

Corollary 1. *The $\exists^*\forall^*$-theory of the class of all finite lattices is $\exists^*\forall^*$-hereditarily undecidable.*

[9] proved that every lattice with only countably many compact elements is isomorpic to an interval in $\mathcal{L}_\Delta$, provided Δ is a large signature. In particular, every finite lattice is embeddable as an interval in $\mathcal{L}_\Delta$.

Our first theorem asserts that for large signatures the lattice of equational theories is $\forall^*\exists^*\forall^*$-hereditarily undecidable. The proof combines the theorems of Schmerl and Ježek in a manner familiar, say, from the work of Burris and McKenzie.

Theorem 1. *Let Δ be a large signature. The $\forall^*\exists^*\forall^*$-theory of $\mathcal{L}_\Delta$ is $\forall^*\exists^*\forall^*$-hereditarily undecidable.*

There is some chance that the methods that led to the conclusion above can be improved to lower the quantifier complexity. While the method of interpretability with parameters used in our proof adds one universal quantifier, the source of most of the complexity lies in the proof of Schmerl's theorem, which depends in turn on a similar theorem [6] for finite graphs. This, in turn, relies ultimately on the transformation of two recursively inseparable sets of numbers into the $\exists^*\forall^*$ setting in a language supplied with a large (but finite) number of relation symbols of ranks no more than 3. It is a classical result from [1] that for a language without operation symbols the set of $\forall^*\exists^*$-validities is decidable. So recursive inseparability results for languages without operation symbols cannot have quanitifier complexity less than $\exists^*\forall^*$. But it is reasonable to expect that in languages supplied with operation symbols rather than relation symbols the complexity of terms involved in equations can be used to off-set the quantifier complexity.

The hereditary undecidability inherent in this approach goes beyond the simple undecidability boundary at issue in our study.

3   The approach via base-undecidability

Jaroslav Ježek, in a series of four papers published in the first half of the 1980's, has extensively developed the theory of definable subsets of

$\mathcal{L}_\Delta$. Among other things, Ježek proved that $\mathcal{L}_\Delta$ has no nonobvious automorphisms and that there is a effective procedure for associating with each finite set Γ of equations of signature Δ a formula $\theta_\Gamma(x)$ that defines in $\mathcal{L}_\Delta$ the orbit, under the action of the automomorphism group, of the equational theory based on Γ. Therefore, we observe that

Γ is equationally inconsistent $\Leftrightarrow \forall x \forall y [\theta_\Gamma(x) \rightarrow y \leq x]$ holds in $\mathcal{L}_\Delta$.

Here we understand that a set Γ of equations is *equationally inconsistent* provided every equation of signature Δ is a logical consequence of Γ, or, what is the same, that the only models of Γ are the one element algebras. Peter Perkins, in his 1966 dissertation, see also [16], proved that there is no algorithm to decide whether a finite set Γ of equations is equationally consistent, provided Δ is a finite (or even recursive) signature that is large. This provides us with a second line of reasoning leading to the conclusion that the elementary theory of $\mathcal{L}_\Delta$ is undecidable, provided Δ is a large signature. An understanding of the quantifier complexity of the formula $\theta_\Gamma(x)$, would lead to a sharper understanding of the quantifier complexity needed to support this undecidability result. A close reading of Ježek's work reveals that while the formula $\theta_\Gamma(x)$ is quite involved, its quantifier complexity depends heavily on neither the signature Δ nor on the set Γ. In particular, the quantifier complexity of $\theta_\Gamma(x)$ is bounded. Unfortunately, the bound established by Ježek's work exceeds the bound we established in the previous section. Nevertheless, it is reasonable to expect that refinements of Ježek's methods, in the restricted case of finite equational bases of the largest equational theory, may well give simpler formulas in place of $\theta_\Gamma(x)$. We also note that Ježek expresses his formulas in the language of lattice-ordered sets (where $\leq$ is the only nonlogical symbol) rather than in the language of lattices. Using operation symbols may not shorten the formulas involved, but it may reduce their quantifier complexity.

One key property of the largest equational theory at play in the reasoning above is that it is a fixed point of every automoprhism of $\mathcal{L}_\Delta$. The other key property of the largest equational theory is its base-undecidability—the result established by Perkins. The work of Perkins was substantially generalized around 1970 by Murskiĭ and the author, see [15,13]. In particular, it was shown that any finitely based equational theory in a large signature which has an equation of the form $t \approx x$ where x is a variable and t is a term in which at least two distinct unary operation symbols occur or else in which some operation symbol of rank at least 2 occurs must be base undecidable. Because Ježek has shown that the automoprhisms of $\mathcal{L}_\Delta$ cannot be very complicated, it is clear that, in a large signature, there are many finite sets Σ so that the equational theory based on Σ is base undecidable and fixed by each automorphism of $\mathcal{L}_\Delta$. Evidently, for such a Σ,

$$\Gamma \text{ and } \Sigma \text{ are bases of the same equational theory}$$

$$\Updownarrow$$

$$\forall x[\theta_\Gamma(x) \leftrightarrow \theta_\Sigma(x)] \text{ holds in } \mathcal{L}_\Delta.$$

This also establishes that the elementary theory of $\mathcal{L}_\Delta$ is undecidable. Once more the quantifier complexity needed to support this undecidability depends of the complexity of θ_Γ andf θ_Σ. There is some prospect that an appropriate choice of the finite set Σ may lead to simpler forms of θ.

4   The approach via undecidable equational theories

In every large signature there are finitely based equational theories that are undecidable. This result probably first appeared in print in a posthumous paper [11] of Mal'tsev, but it was known in principal as early as 1947 in the works [17] and [12]. Perhaps the most surprising result of this kind in the 1976 theorem of Freese [7] that the equational theory of modular lattices is undecidable. (He even proves the undecidability of the set of equations true in all modular lattices that involve no more than 5 variables—a result improved by Christian Herrmann to 4 variable equations, see [8].). For our purposes, the interesting thing about the equational theory of modular lattices is that it is fixed by every automorphism of $\mathcal{L}_\Delta$, where Δ is the signature of lattices. Let Σ be a finite equational base for the theory of modular lattices and let $s \approx t$ be any equation in the language lattices. Then

$$\Sigma \vdash s \approx t \Leftrightarrow \mathcal{L}_\Delta \models \forall x \forall y[\theta_{s \approx t}(x) \wedge \theta_\Sigma(y) \to x \leq y].$$

In this way we can once again conclude that the elementary theory of $\mathcal{L}_\Delta$ is undecidable, at least when Δ is the signature of lattices. Of course, the quantifier complexity needed to support this undecidability depends on the complexity of θ. To make this line of reasoning apply to all large signature we have to construct, for each such signature a finite set Σ of equations so that the equational theory based on Σ is both undecidable and also fixed under every automorphism of $\mathcal{L}_\Delta$.

References

1. Bernays, P., Schönfinkel, M.: Zum Entscheidungsproblem der mathematischen Logik. Math. Ann. **99**(1) (1928) 342–372 68
2. Burris, S., McKenzie, R.: Decidable varieties with modular congruence lattices. Bull. Amer. Math. Soc. (N.S.) **4**(3) (1981) 350–352 67
3. Burris, S.: On the structure of the lattice of equational classes $\mathcal{L}(\tau)$. Algebra Universalis **1**(1) (1971) 39–45 66
4. Burris, S., Sankappanavar, H.P.: Lattice-theoretic decision problems in universal algebra. Algebra Universalis **5**(2) (1975) 163–177 66
5. Eršov, J.L., Lavrov, I.A., Taĭ manov, A.D., Taĭ clin, M.A.: Elementary theories. Uspehi Mat. Nauk **20**(4 (124)) (1965) 37–108 67
6. Eršov, J.L., Taĭ clin, M.A.: Undecidability of certain theories. Algebra i Logika Sem. **2**(5) (1963) 37–41 68
7. Freese, R.: Free modular lattices. Trans. Amer. Math. Soc. **261**(1) (1980) 81–91 70
8. Herrmann, C.: On the word problem for the modular lattice with four free generators. Math. Ann. **265**(4) (1983) 513–527 70
9. Ježek, J.: Intervals in the lattice of varieties. Algebra Universalis **6**(2) (1976) 147–158 68
10. Lerman, M.: Degrees of unsolvability. Perspectives in Mathematical Logic. Springer-Verlag, Berlin (1983) Local and global theory. 67
11. Mal'tsev, A.I.: Identity relations on manifolds of quasigroups. Mat. Sb. (N.S.) **69 (111)** (1966) 3–12 70
12. Markoff, A.: On the impossibility of certain algorithms in the theory of associative systems. C. R. (Doklady) Acad. Sci. URSS (N.S.) **55** (1947) 583–586 70
13. McNulty, G.F.: The decision problem for equational bases of algebras. Ann. Math. Logic **10**(3-4) (1976) 193–259 69
14. Monk, J.D.: Mathematical logic. Springer-Verlag, New York-Heidelberg (1976) Graduate Texts in Mathematics, No. 37. 67
15. Murskiĭ, V.L.: Unrecognizable properties of finite systems of identity relations. Dokl. Akad. Nauk SSSR **196** (1971) 520–522 69
16. Perkins, P.: Unsolvable problems for equational theories. Notre Dame J. Formal Logic **8** (1967) 175–185 69
17. Post, E.L.: Recursive unsolvability of a problem of Thue. J. Symbolic Logic **12** (1947) 1–11 70
18. Rabin, M.O.: A simple method for undecidability proofs and some applications. In: Logic, Methodology and Philos. Sci. (Proc. 1964 Internat. Congr.). North-Holland, Amsterdam (1965) 58–68 67
19. Sachs, D.: Identities in finite partition lattices. Proc. Amer. Math. Soc. **12** (1961) 944–945 66
20. Skolem, T.: Logisch-kombinatorische untersuchungen über die erfüllbarkeit und beweisbarkeit mathematischen sätze nebst einem theoreme über dichte mengen. Vidensskapsselskapets skrifter I, Matematisk-natruv. klasse, Videnskabsakademiet i Kristiania **4** (1920) 1–36 66

21. Skolem, T.: Selected works in logic. Edited by Jens Erik Fenstad. Universitetsforlaget, Oslo (1970) 66
22. Tarski, A.: Undecidable theories. Studies in Logic and the Foundations of Mathematics. North-Holland Publishing Company, Amsterdam (1953) In collaboration with Andrzej Mostowski and Raphael M. Robinson. 67
23. Whitman, P.M.: Free lattices. Ann. of Math. (2) **42** (1941) 325–330 66

On nonblocking words

Irina Mel'nichuk*

Omsk Humanitarian Academy, Russian Federation
zakina1@rambler.ru

A word u is called *blocking* [8] if for any finite alphabet the set of words of that alphabet which do not contain images of the word u is finite. Recall that an image of the word u is the result of substituting words in place of the letters in the word u (note that the same letters are replaced by the same words). A nonblocking word u is called *n-avoidable* if on some n-letter alphabet (and therefore on all n-letter alphabets) there exists infinitely many words that do not contain images of u. In Problem 3 from [3] the question of finding for any nonblocking word the smallest alphabet on which the word is avoidable is formulated. Some partial results related to this question are known.

It is shown in [1,7] that the word x^2 is avoidable on the 3-letter alphabet and that x^3 is avoidable on the 2-letter alphabet. It is established in [4] that the set of words of the form

$$x_1 \ldots x_n x_{f(1)} \ldots x_{f(n)},$$

where f is a permutation of the numbers $1, \ldots, n$, is avoidable on the 7-letter alphabet. In [5] it is shown that the set of all doubled words on the n-letter alphabet is avoidable on an alphabet with $3\lfloor \frac{n}{2} \rfloor + 3$, where $\lfloor \frac{n}{2} \rfloor$ is the integer part of $\frac{n}{2}$. In [6], infinitely many words of the 4-letter alphabet that avoid every complete word are constructed. It is also shown in [6] that this cannot be done with an alphabet consisting of 3 letters. Theorems 1.2 and 1.3 from [2] give upper bounds on the cardinality of alphabets in which any nonblocking word is avoidable.

Let u be a nonblocking word and let the number of distinct letters in u be denoted by $\alpha = \alpha(u)$. Denote by $m = m(u)$ the minimal number n such that u is n-avoidable. By Theorem 1.2 in [2] for any nonblocking word u, we have $m < 4(\alpha + 2)\lceil \log(\alpha + 2) \rceil$. By Theorem 1.3 in [2], $m < 9\alpha + 20$, for any nonblocking word u. The goal of this paper is the following improvement of the bound on m.

*I am very grateful to Professor Kira Adaricheva, who found me—at the request of Professor George McNulty—though it was not easy, and who advised me to put this article on the arXiv. Special thanks to Professor McNulty, who translated the article from the Russian language and encouraged me to return to this paper, which I wrote in 1996 but never published.

Theorem 1. *For any nonblocking word u, we have $m(u) \leq 2\alpha(u) + 4$.*

Proof. Denote by k the number out of the two numbers $2\alpha(u) + 2$ and $2\alpha(u) + 4$ which is divisible by 4. We prove that u is k-avoidable. We assume that $\alpha(u) > 1$, becasue in case $\alpha(u) = 1$ the state in the theorem follows from [7].

We need the description of blocking word obtained in [8]. Consider the words over the alphabet $\Xi = \{\xi_1, \xi_2, \dots\}$. Let $z_1 = \xi_1$ and by induction $z_{n+1} = z_n \xi_{n+1} z_n$, where $n = 1, 2, \dots$. Denote by $F_1(A)$ the set of all nonempty words over the alphabet A and denote by $\alpha(u)$ the set of all letters of the word u.

For the word u the mapping $f: \alpha(u) \to F_1(\Xi)$ such that $f(u)$ is a subword of z_n for some n, is called a B-mapping of the word u. A. I. Zimin in [8] proved that a word is blocking if and only if there exists a B-mapping of this word.

We consider in the symmetric group S_k the following permutations:

$$f_1 = (1, 3, 5, \dots, k-1) \qquad\qquad g_1 = (2, 6, \dots, k-2)$$
$$f_2 = f_1^2 \qquad\qquad\qquad\qquad g_2 = g_1 f_1$$

$$\vdots \qquad\qquad\qquad\qquad\qquad \vdots$$

$$f_{\frac{k}{2}} = f_1^{\frac{k}{2}} \qquad\qquad\qquad g_{\frac{k}{2}} = g_1 f_1^{\frac{k}{2}-1}.$$

We obtain k different permutations and we denote them by $v_1, \dots, v_k$. For the selected words, we fix k words $a_1, \dots, a_k$ over the alphabet $X = \{x_1, \dots, x_k\}$, setting $a_i = x_{v_i(1)} \dots x_{v_i(k)}$ for each i.

Definition 1. *We define the mapping $\varphi_0: X \to F_1(X)$ by the rule $\varphi_0(x_i) = a_i$. We denote by $J_1 = \varphi_0(x_1)$. Suppose that J_{m-1} is already defined; then we set $J_m = \varphi_0(J_{m-1})$.*

To prove the theorem it is sufficient to show that for $m \geq 1$ the word J_m does not contain images of the word u. The proof of this fact follows from the following proposition.

Proposition 1. *Let u be any word with no more than $\frac{k}{2} - 1$ distinct letters. Let $\varphi: \alpha(u) \to F_1(X)$ be any mapping so that the word $\varphi(u)$ is a subword of J_n for some $n \geq 1$. Then there is a B-mapping of the word u, that is the word u is blocking.*

Proof (Proof of Proposition 1). We consider the set $A = \{a_1, a_2, \dots, a_k\}$ as an alphabet. Then for $n \geq 1$ the word J_n can be construed as a word over the alphabet A; let's call it a *chain* in this case. An ocurrence of b from $F_1(A)$ in the chain J_n will be called a subchain b; in particular, the occurrence of a_i will be called the *link* a_i: thus, the subchain a_i in J_n is an occurrence of a word of length 1 over A, whereas a_i is a word of length k in $F_1(X)$. We shall call the word a of $F_1(X)$ a subword of the subchain b, if the corresponding word b in the alphabet X contains the subword a.

Let there be given an occurrence of the word a in J_n. A *closure* of the occurrence a is the smallest subchain of the chain J_n, containing the given occurrence a. We denote the closure of a by $[a]$.

The proof of the proposition will be carried out by induction on the length $[\varphi(u)]$

It is clear that if $[\varphi(u)]$ coincides with one of the letters a_i, for $i \in \{1, \ldots, k\}$, then u is a blocking word.

Induction step. Suppose that $\varphi(u)$ is a subword of J_n for some $n \geq 1$, and the sub-chain $[\varphi(u)]$ contains more than one link. Let us prove, using the induction premise, that there exists a B-mapping of u.

Definition 2. *A word b of $F_1(X)$ is said to be* basic *if it is a subword of a_i for some $i \in \{1, \ldots, k\}$ and contains at least two letters of X with even indices.*

Definition 3. *If $a_i = b_1 b_2$ and $a_j = c_1 c_2$, where $b_2, c_1 \in F_1(X)$, then the word $b_2 c_1$ is said to be* adjacent.

Definition 4. *The subword $b = cd$, $c \neq \wedge$, $d \neq \wedge$ of the word a_i, $i \in \{1, \ldots, k\}$ is said to be* broken *if for some occurrence of the letter y in u the corresponding closure $\varphi(u)$ adheres to the last (first) link a_i, and c (d) is the right (left) subword $\varphi(y)$.*

Definition 5. *For every r^{th} occurrence of the letter y in the word u, we fix the sub-$[\varphi(y)]$, which is the closure of the rth occurrence of $\varphi(y)$ in J_n corresponding to the rth entry of y in u. Suppose that $[\varphi(y)]_r = c\varphi(y)d$. If $c \neq \wedge$, then the first link $[\varphi(y)]_r$ is called* broken. *If $d \neq \wedge$, then the last link will be called broken, the other links $[\varphi(y)]_r$ will be undefined. For the sake of brevity, $[\varphi(y)]_1$, we denote $[\varphi(y)]$.*

Definition 6. *The letter x with an even index a_i is called* basic *if there is no such occurrence of the letter y of the word u that the link a_i is the first for the corresponding closure $\varphi(y)$ of the word $\varphi(y)$ contains x to the first or second letter.*

It follows from the definition that if x is a basic letter of a_i, the link a_i is the first in the chain $[\varphi(y)]_r$, then either the occurrence of a basic word of the form $x_i x_j x$ in a_i is an occurrence in $\varphi(y)$, or the occurrence of x in a_i is not an occurrence in $\varphi(y)$.

Lemma 1. *The words $a_1, \ldots, a_k$ have the following properties:*

   a) *a basic word is a subword of a_i for a single i of the set $\{1, \ldots, k\}$;*
   b) *if b is a left subword of a_i and c is a left subword of a_j and $b = c$, $|b| \geq 2$, then $a_i = a_j$;*
   c) *there is no adjacent word that is a subfactor a_i for some $i \in \{1, \ldots, k\}$;*
   d) *the word J_n has no equal adjacent subwords;*
   e) *each a_i has a basic letter.*

Definition 7. *For each word a_i, $i \in \{1, \ldots, k\}$ we fix the basic letter a_i and call it the major basis letter a_i.*

Definition 8. *We fix the mapping $C : F_1(X) \rightarrow F_1(\Xi)$ in the following way. We represent the natural number p in the form $p = 2^i + r2^{i+1}$ for some integers $r \geq 0, i \geq 0$. This view is unambiguous. We set $C(x_p) = \xi_{i+1}$ for any x_p, where $1 \leq p \leq k-1$, $p = 2^i + r2^{i+1}$. Then the word $C(x_1 x_2 \ldots x_{k-1})$ coincides for some number t with the left subword Z_t. Choose the smallest such t. Suppose that $Z_t = C(x_1 x_2 \ldots x_{k-1})Z'$, then we set $C(x_k) = Z'\xi_{t+1}$.*

By definition, $C(x_1) = C(x_3) = \ldots = C(x_{k-1}) = \xi_1$, $C(x_2) = C(x_6) = \ldots = \xi_2$, $C(x_1 x_2 \ldots x_k) = Z_t \xi_{t+1}$. Since the word Z_t does not change from the permutation of the letters ξ_1 or ξ_2, for any a_i, $i \in \{1, \ldots, k\}$, $C(a_i) = Z_t \xi_{t+1}$.

Lemma 2. *Suppose that for every y in $\alpha(u)$ the word $\varphi(y)$ satisfies the condition: if a_i is a subword $[\varphi(y)]$, then the occurrence of the major basis letter a_i is not an occurrence in $\varphi(y)$. Then u is a blocking one.*

Proof. Let the mapping φ satisfy the condition of the lemma. In this case, each of the words a_i, $i \in \{1, \ldots, k\}$ is not a subword of $\varphi(y)$, since by Lemma 1 e), each a_i has a major basis letter. This means that $\varphi(y)$ is for some i, j the subword $a_i a_j$, and at least the last letter a_j - the letter x_k does not enter $\varphi(y)$. Therefore, $C(\varphi(y))$ is a subword of $Z_t \xi_{t+1} Z_t$, that is, it is blocking. Lemma 2 is proved.

We assume in what follows that the condition of Lemma 2 does not hold, that is, for some y in $\alpha(u)$ the word $\varphi(y)$ contains the occurrence of the major basis letter of some a_i from $[\varphi(u)]$. It follows from Definition 7 of the major basis letter of the word that if for a certain p-th occurrence of y in u the chain $[\varphi(y)]_p$ begins (ends) with a link a_i, and the occurrence of the major basis letter of the word a_i is an occurrence of $\varphi(y)$, then for any r-th occurrence of y in u the chain $[\varphi(y)]_r$ begins (ends) with the subword a_i.

Definition 9. *We associate the word u with the word u_1 in the following way: delete from u occurrences of all letters y for which $[\varphi(u)]$ consists of broken links (and therefore contains one or two links), and the occurrence of the major basis letter of each of these links is not an occurrence in $\varphi(y)$.*

On the set P of all links of the chain $[\varphi(u)]$, we construct a function f : $P \rightarrow \{R, L, 0\}$. Suppose that for some r-th occurrence of x in u the sub-chain $[\varphi(x)]_r$ contains the link a_i.

a) *If a_i is undefined (Definition 5), or $[\varphi(x)]_r = a_i$ for x from $\alpha(u_1)$, then we set $f(a_i) = 0$.*

b) *the broken link a_i is the first of the sub-chain $[\varphi(x)]_r$, $\varphi(x) = sb$ where s is the right-hand subword a_i. If the major basis letter a_i enters s, then we set $f(a_i) = R$;*

c) *the broken link a_i is the last one in the sub-chain $[\varphi(x)]_r$, $\varphi(x) = bs$ where s is the left subword a_i. If the major basis letter a_i enters s, then we set $f(a_i) = L$;*

d) *suppose that for the broken link a_i - the conditions of points a)–c) are not fulfilled, that is, the function f is not defined yet, then we set $f(a_i) = 0$. It is obvious that this case is possible when a_i is the last link in the chain $[\varphi(u)]$ and the major basis letter a_i does not enter $\varphi(u)$.*

For each broken link, there is a single major basis letter, so the definition of the function is correct.

Comment For any r-occurrence of y in u_1, the value of f for the last (first) link is $[\varphi(y)]_r$ is equal to the value of f for the last (first) link $[\varphi(y)]$.

Indeed, assume that $y \in \alpha(u_1)$, $[\varphi(y)]_r = b\, a_j$, $[\varphi(y)] = d\, a_i$. Let $f(a_i) = L$, $a_i = a_{i_1} a_{i_2}$, where a_{i_1} is the right subword of $\varphi(y)$. Then, by definition of f, the major basis letter a_i enters a_{i_1}. Since the major basis letter has an even index, the length of a_{i_1} is greater than or equal to two, and by Lemma 1 b) $a_i = a_j$ and $f(a_i) = L = f(a_j) = L$.

Let $f(a_i) = R$, $[\varphi(y)]_r = a_j\, b$, $[\varphi(y)] = a_l\, d$, $a_i = a_{i_1} a_{i_2}$, where a_{i_2} is the left subword $\varphi(y)$. Then the major basis letter a_i is contained in a_{i_2}, and hence a_{i_2} contains a basic subword that uniquely determines the word a_i and $a_i = a_j$.

If $f(a_i) = 0$ and a_i is undefined (Definition 5), or $[\varphi(y)]_r = a_i$ for y from $\alpha(u_1)$, then $[\varphi(y)]$ contains the basic subword a_i, and therefore, for any r-th occurrence of y in u_1, $[\varphi(y)]_r = a_i$.

We construct the map $\Psi : \alpha(u_1) \rightarrow F_1(A)$. Let $y \in \alpha(u_1)$.

a) Suppose that $[\varphi(y)] = a_i$. We set $\Psi(y) = a_i$.

b) Suppose that the sub-chain $[\varphi(y)]$ contains more than one link, the first link of this sub-chain is a_i, the last a_j, that is, $[\varphi(y)] = a_i Ba_j$. We assume

$$
\Psi(y) = \begin{cases}
a_i Ba_j, & \text{if } f(a_i) = R, f(a_j) = L, \\
a_i B, & \text{if } f(a_i) = R, f(a_j) \neq L, \\
Ba_j, & \text{if } f(a_i) \neq R, f(a_j) = L, \\
B, & \text{if } f(a_i) \neq R, f(a_j) \neq L.
\end{cases}
$$

Since every broken or unbroken link from the sub-chain $[\varphi(y)]$, except, may be the first and the last, it enters $[\Psi(y)]_r$ for single $r \geq 1$, y from $\alpha(u_1)$, then $\Psi(u_1)$ is a subword of J_n and by the definition of the mapping Ψ, for every y in $\alpha(u_1)$, $\Psi(y) = [\Psi(y)]$.

By displaying Ψ, we construct the map $\Psi_1 : \alpha(u_1) \to F_1(A)$, assuming that if $\Psi(y) = a_{i_1} \ldots a_{i_m}$, then $\Psi_1(y) = x_{i_1} \ldots x_{i_m}$.

Then $\Psi_1(y)$ is a subword of J_{n-1} and the length of $[\Psi_1(u_1)]$ is less than the length $[\Psi(u)]$, therefore, by the inductive hypothesis, u_1 is a blocking word and there exists a B-mapping of the word u_1.

Let $G(u_1)$ be the subword of Z_r. We construct the map $G_1 : \alpha(u) \to F_1(\Xi)$ and prove that this is a B-map of the word u_1.

We introduce the mapping $H : \Xi \to F_1(\Xi)$ by setting $H(\xi_i) = Z_t \xi_{i+t+1} Z_t \xi_{t+1}$ (t is taken from Definition 8). It is easy to see that $H(Z_r)$ is a subword of Z_{r+t+1} and therefore $H(G(u_1))$ is a subword of Z_{r+t+1}.

Definition 10. *For each $y \notin \alpha(u_1)$ we set $G_1(y) = C(\varphi(y))$ (the map C is given in definition 8).*

Let $y \in \alpha(u_1)$.

a) *$\varphi(y)$ is a subword of a_i for some $i \in \{1, \ldots, k\}$. Suppose that $a_i = m\varphi(y)s$, $H(G(y)) = Z_t b Z_t \xi_{t+1}$. Then we set $G_1(y) = pbC(m\varphi(y))$, where p denotes the word obtained from the word $C(\varphi(y)s)$ by discarding the last letter ξ_{t+1}. Note that $H(G(y))$ is a subword of Z_{r+t+1} and $G_1(y)$ is the subword of $H(G(y))$.*

b) *Let the chain $[\varphi(y)]$ contain more than one link, the first link of this sub-chain a_i, the last a_j, that is, $[\varphi(y)] = a_i Ba_j$, $\varphi(y) = mBs$. Suppose that $H(G((y)) = Z_t b Z_t \xi_{t+1}$. We denote by p the word obtained from $C(m)$ by discarding the last letter xi_{t+1}. We believe:*

$$
G_1(x) = \begin{cases}
pbC(s), & \text{if } f(a_i) = R, f(a_j) = L, \\
pbZ_t\xi_{t+1}C(s), & \text{if } f(a_i) = R, f(a_j) \neq L, \\
C(m)Z_t bC(s), & \text{if } f(a_i) \neq R, f(a_j) = L, \\
C(m)Z_t b Z_t\xi_{t+1}C(s), & \text{if } f(a_i) \neq R, f(a_j) \neq L.
\end{cases}
$$

Lemma 3. *For the constructed map* $G_1 : \alpha(u) \to F_1(\Xi)$, *the word* $G_1(u)$ *is the subword* $Z_t \xi_{t+1} H(G(u_1)) Z_t$.

Proof. We prove the lemma by induction on the length of the word u_1. By Lemma 2 the word u_1 is not empty. Let u_1 be the one-letter word x, $H(G(x)) = Z_t b_t Z_t \xi_{t+1}$, and u_1 be obtained from the word $u = dxh$ by deleting the subwords d, h. Each of the chains $[\varphi(d)]$, $[\varphi(h)]$ contains no more than two links. To be specific, let $[\varphi(d)] = a_i$ and $[\varphi(h)] = a_j a_m$ (other cases are treated similarly). We write a_i, $a_j a_m$ in the form $a_i = k_1 \varphi(d) k_2$, $a_j a_m = r_1 \varphi(h) r_2$. Since the letters of the subword h were deleted from the word u, then $\varphi(h)$ does not contain the occurrence of the major basis base letter a_j and does not contain the occurrence of the major basis letter a_m. Suppose that $f(a_i) \neq R$ (we similarly consider the case when $f(a_i) = R$). Since the link a_j is not the last in the chain $[\varphi(u)]$ and the letters of the word h are deleted from u, then $f(a_j) = L$ and $f(a_m) = 0$.

By the Definition 10, $G_1(x) = C(k_2) Z_t b C(r_1)$, $G_1(d) = C(\varphi(d))$, $G_1(h) = C(\varphi(h))$. Therefore, $G_1(dxh) = C(\varphi(d)) C(k_2) Z_t b C(r_1) C(\varphi(h))$, where $C(\varphi(d)) C(k_2)$ is the right subword of $C(a_i) = Z_t \xi_{t+1}$, and $C(r_1) C(\varphi(h))$ is the left subword of $Z_t \xi_{t+1} Z_t$ containing $Z_t \xi_{t+1}$. The induction base is proved.

Suppose that the word u_1 has the form $u_1 = s_1 xyw_1$, where $x, y \in \alpha(u_1)$ and u_1 is obtained from the word $u = sxvyw$ by deleting the by-word v and some other letters from the words s, w. Let $G(u_1) = b_1 b_2$ be the subword of Z_r, where $G(s_1 x) = b_1$, $G(yw_1) = b_2$. By the induction hypothesis, $G_1(sx)$ is the subword $Z_t \xi_{t+1} H(G(s_1 x)) Z_t$ and $G_1(yw)$ is the subword $Z_t \xi_{t+1} H(G(yw_1)) Z_t$. The letters of the word v are deleted from u when getting u_1, therefore $[\varphi(v)]$ contains no more than two links. Suppose that $[\varphi(v)] = a_i$ (the case where $[\varphi(v)]$ is empty or contains two links, is treated similarly). Write the word a_i in the form: $a_i = k_1 [\varphi(v)] k_2$. Let $H(G(s_1 x)) = Z_t b_1 Z_t \xi_{t+1}$, $H(G(yw_1)) = Z_t b_2 Z_t \xi_{t+1}$.

a) Suppose that $f(a_i) = R$. Then by Def. 10, $G_1(sx) = d_1 b_1 Z_t \xi_{t+1} C(k_1)$, $G_1(yw) - pb_2 d_2$ where p is the word obtained from $C(k_2)$ (Def. 8) by discarding the last letter ξ_{t+1} and d_1, d_2— the subword Z_{t+1}. But $C(k_1) C(\varphi(v)) p = Z_t$, and therefore, $G_1(sxvyw) = d_1 b_1 Z_{t+1} b_2 d_2$.

b) Suppose that $f(a_i) = L$. Then, by Def. 10, $G_1(sx) = d_1 b_1 C(k_1)$, $G_1(yw) = C(k_2) Z_t b_2 d_2$. But $C(k_1) [\varphi(v)] k_2) = Z_t \xi_{t+1}$ therefore $G_1(sxvyw) = d_1 b_1 Z_{t+1} b_2 d_2$.

c) Suppose that $f(a_i) = 0$. Since a_i is a broken link that is not extreme in the chain $[\Psi_1(u_1)]$, then either $[\varphi(x)] = a_i$, or $[\varphi(y)] = a_i$. If $[\varphi(x)] = a_i$, the proof is the same as in case b). If $[\varphi(y)] = a_i$, then the proof is similar to the case a).

Lemma 3 is proved.

Thus, the map $G_1 : \alpha(u) \to F_1(\Xi)$ is a B-map of u, and the proposition is proved.

It follows that if u is a non-blocking word with the number of different letters $\alpha(u)$, then the infinite set of words J_m, $m = 1, 2, \ldots$ does not contain the values of the word u, that is, u is avoided in the alphabet containing $2\alpha(u) + 4$ letters.

The theorem is proved.

References

1. Arshon, E.S.: Proof of the existence of n-valued infinite asymmetric sequences. Mat. Sb. (1939) 769–779 73
2. Baker, K.A., McNulty, G.F., Taylor, W.: Growth problems for avoidable words. Theoret. Comput. Sci. **69**(3) (1989) 319–345 73
3. Bean, D.R., Ehrenfeucht, A., McNulty, G.F.: Avoidable patterns in strings of symbols. Pacific J. Math. **85**(2) (1979) 261–294 73
4. Evdokimov, A.A.: Strongly asymmetric sequences generated by a finite number of symbols. Dokl. Akad. Nauk SSSR **179** (1968) 1268–1271 73
5. Mel'nichuk, I.L.: Existence of infinite finitely generated free semigroups in certain varieties of semigroups. Leningrad. Gos. Ped. Inst., Leningrad (1985) 146 73
6. Petrov, A.N.: A sequence that avoids every complete word. Mat. Zametki **44**(4) (1988) 517–522, 558 73
7. Thue, A.: Über unendliche Zeichenreihen. Universitetsforlaget, Oslo (1977) 139–158 First published: 1906, in Mat.-Nat. Kl. Christiana, , Vol. 7, P. 1-22. 73, 74
8. Zimin, A.I.: Blocking sets of terms. Mat. Sb. (N.S.) **119(161)**(3) (1982) 363–375, 447 73, 74

Extending partial projective planes

J. B. Nation

Department of Mathematics, University of Hawai'i, Honolulu, HI 96822
jb@math.hawaii.edu

Keywords: finite projective plane, partial projective plane

Abstract. This note discusses a computational method for constructing finite projective planes.

There are a number of interesting problems concerning finite non-desarguean projective planes. One would hope that these problems would admit an algebraic, or geometric, or combinatorial solution. But it may just be that the existence, or non-existence, of certain types of planes is an accident of nature. With that in mind, since 1999 the author has been trying various computer programs to construct non-desarguean projective planes. While all these attempts have failed, hope springs eternal, and this note describes a set of problems and some ideas for addressing them.

This note is based on a talk given to the Courant Institute geometry seminar in October 2017. The author appreciates the hospitality and encouragement from the participants of the seminar.

1 Basics

Recall that a *projective plane* is an incidence structure of points and lines satisfying these axioms

- Two points determine a unique line.
- Two lines intersect in a unique point.
- There exist four points with no three on a line.

Each finite plane has an *order n* such that there are

- $n + 1$ points on each line,
- $n + 1$ lines through each point,
- $n^2 + n + 1$ total points,

- $n^2 + n + 1$ total lines.

An elementary construction allows us to construct a projective plane starting from the affine plane over any division ring. Indeed, every projective plane can be coordinatized by a ternary ring [11]. For the basic combinatorics and coordinatization of projective planes, see [1,6,12].

Desargues' Law, a property that holds in some projective planes and not others, is illustrated in Figure 1 and explained in the caption. Planes that satisfy Desargues' Law are called *desarguean*. A projective plane is desarguean if and only if it can be coordinatized by a division ring. Thus from finite fields we obtain projective planes of order q for any prime power $q > 1$. The same construction yields non-desarguean projective planes coordinatized by various quasi-fields; these are of prime power order $q \geq 9$.

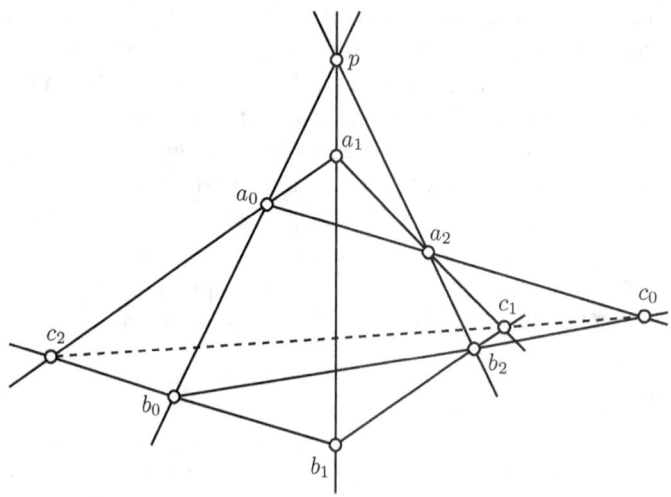

Fig. 1. Desargues' Law: If the lines $a_0 \vee b_0$, $a_1 \vee b_1$, and $a_2 \vee b_2$ intersect in a point p, then the points c_0, c_1, and c_2 are colinear, where $c_i = (a_j \vee a_k) \wedge (b_j \vee b_k)$ for $\{i, j, k\} = \{0, 1, 2\}$. In a failure of Desargues' Law, the lines $c_0 \vee c_1$, $c_0 \vee c_2$, and $c_1 \vee c_2$ are distinct.

There are four isomorphism types of planes of order 9, including the one coordinatized by a field of order 9, and a non-desarguean plane coordinatized by a Hall quasi-field, and its dual. The fourth type, the Hughes plane, admits no such nice description.

A classic result of Bruck and Ryser [3] shows that some orders are impossible: *If $n \equiv 1$ or $2 \bmod 4$ and there is a plane of order n, then n is a sum of two squares.* Beyond the Bruck-Ryser Theorem, only one more restriction is known: Lam, Theil and Swiercz [13] proved that there is no plane of order 10.

That leaves the existence of a projective plane of the following orders unknown: 12, 15, 18, 20, 24, 26, 28, ...

A subplane of a finite projective plane need not have order dividing the order of the plane. Indeed, H. Neumann showed that every Hall plane has a subplane of order 2 (see [10]).

2 Partial projective planes

A *partial projective plane* is a collection of points and lines, and an incidence relation, so that

- two points lie on at most one line,
- two lines intersect in at most one point.

M. Hall showed that every finite partial plane can be extended to a projective plane (usually infinite) [11]. In retrospect, it is not hard to see how to build this free extension.

Now projective planes correspond to simple, complemented, modular lattices of height 3. To form a projective plane from such a lattice, take the points to be the elements of height 1, and the lines to be the elements of height 2. Upper semimodularity means that 2 points join to a unique line, while lower semimodularity means that 2 lines meet in a unique point. Simplicity guarantees that there are 4 points in general position.

This suggests that we employ partial planes that are meet semilattices. A *semiplane* is a collection of lines and points, with an incidence relation, such that any two lines intersect in a unique point. A canonical example of a semiplane is formed by taking any subset of the lines of a plane, together with the points that are intersections of those lines.

3 Four questions

That brings us to four basic questions about finite projective planes.

- Is there a finite projective plane of non-prime-power order (necessarily non-desarguean)?
- Is there a non-desarguean plane of prime order?
- Does every finite non-desarguean plane contain a subplane of order 2?
- Does every finite partial plane have an extension to a finite plane?

If we fix a desired order n, the general plan for extending a finite partial plane to a plane of order n is straightforward enough.

- Start with a semiplane that contains your desired configuration (e.g., a failure of Desargues' Law or a plane of order 2).
- As long as possible add lines, with their intersections with existing lines, one at a time, keeping a semiplane structure and at most $n + 1$ points-per-line.
- Intersections can be new points or old points.
- If you get $n^2 + n + 1$ lines, the semiplane is a plane [5].
- Otherwise, when adding a line is no longer possible, back up and try again.

Did we mention that you should be *patient*, as the program could take tens of thousands of years?

Nonetheless, there is a simple turnaround criterion from [14]. Given n and a semiplane $\Pi = \langle P_0, L_0, \leq_0 \rangle$, define

$$\rho_n(\Pi) = \sum_{\ell \in L_0} r_\Pi(\ell) + n^2 + n + 1 - |P_0| - |L_0|(n + 1)$$

where $r_\Pi(\ell)$ denotes the number of points on the line ℓ in the semiplane Π. If $\Pi = \langle P_0, L_0, \leq_0 \rangle$ can be extended to a projective plane $\Sigma = \langle P, L, \leq \rangle$ of order n, then

$$\rho_n(\Pi) = |\{p \in P : p \not\leq \ell \text{ for all } \ell \in L_0\}|.$$

Hence Π can be extended to a projective plane of order n only if $\rho_n(\Pi) \geq 0$.

There are nice extension theorems for some types of partial structures, summarized in Chapter 9 of Dénes and Keedwell [7], and updated in [8]. We note especially the results of Bruck [2] and Dow [9]. For a logical approach, see Conant and Kruckman [4].

4   Non-desarguean planes

In order to apply this program to construct a non-desarguean plane of questionable order, we must first extend a non-desarguean configuration to a semi-plane. A non-desarguean configuration has 10 points and 12 lines, see Figure 1. To form a semiplane, those lines can intersect in various ways: the intersections could be new points or old ones. The result is a semiplane with 12 lines and between 20 and 37 points. Seffrood proved that there are 875 such non-desarguean semiplanes, which fall into 105 isomorphism classes. For some pairs (A,B) of the 105 types, if a plane contains a semiplane of type A then it contains one of type B. There are 15 non-desarguean semiplanes that are minimal in the sense that every non-desarguean plane must contain one of these 15 semiplanes. Thus a program to construct finite non-desarguean planes can use one of these 15 minimal semiplanes as a starting configuration. (The results in this paragraph are from Seffrood and Nation [14].)

So far, our programs have yielded

- semiplanes of order 11 with 40 lines (a plane has 133 lines),
- semiplanes of order 12 with 44 lines (a plane has 157 lines),
- semiplanes of order 13 with 48 lines (a plane has 183 lines),
- a semiplane of order 15 with 56 lines (a plane has 241 lines).

Most of these semiplanes have the full number of points.

This is not as bad as it first seems. When we tested the program by constructing a Hall plane of order 9, it turned out that once the semiplane had 35 lines, from that point on there was a unique choice for how to extend it with a new line. Thus the program extended the semiplane with 35 lines to a plane with 91 lines in a matter of seconds. Hence we suggest the following problem:

Find $f(n)$ such that every semiplane with at least $f(n)$ lines and at most $n + 1$ points-per-line can be extended to a plane of order n.

There are results of this nature for latin squares. A projective plane of order n can be constructed from a set of $n - 1$ mutually orthogonal $n \times n$ latin squares (MOLS), and *vice versa* [7,8]. Shrikhande [15] proved that, for $n > 4$, any set of $n - 3$ mutually orthogonal $n \times n$ latin squares can be extended to a complete set of $n - 1$ MOLS, and Bruck [2] proved the same result for any collection of $n - 1 - (2n)^{\frac{1}{4}}$ MOLS. (See Chapter 9 of [7].)

5   Other starting configurations

Pappus' Law is illustrated in Figure 2 and explained in the caption. A de-sarguean plane can be coordinatized by a division ring; that division ring is commutative if and only if the plane satisfies Pappus' Law. Since every finite division ring is a field, this suggests that we start with a semiplane generated by a failure of Pappus' Law. A non-pappian configuration has 9 points and 11 lines, so there should be fewer options for completing it to a semiplane.

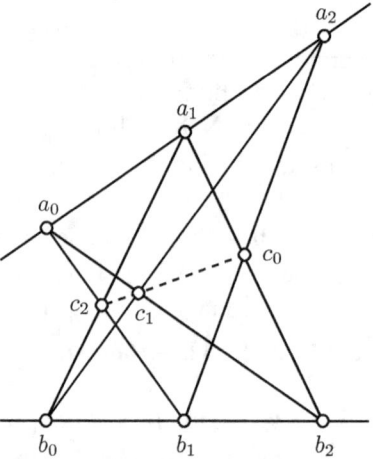

Fig. 2. Pappus' Law: If the points a_0, a_1, and a_2 are colinear, and the points b_0, b_1, and b_2 are colinear, then the points c_0, c_1, and c_2 are colinear, where $c_i = (a_j \vee b_k) \wedge (b_j \vee a_k)$ for $\{i, j, k\} = \{0, 1, 2\}$. In a failure of Pappus' Law, $c_0 \vee c_1$, $c_0 \vee c_2$, and $c_1 \vee c_2$ are three distinct lines.

Hanna Neumann's Fano planes sitting inside Hall planes suggests another idea: *What happens if you start with a plane of order 2 and try to extend it to a plane of order n?* A starting semiplane with a Fano plane and one extra line would have 8 lines and 14 points as the only option.

Freese tried a variation on this theme, with a program to look for intermediate subplanes in a Hall plane, between a Fano subplane and the whole Hall plane [10]. Perhaps it is time to revisit this approach.

A theorem from folklore is that if a plane of order n has a subplane of order $r < n$, then either $n = r^2$ or $n \geq r^2 + r$. Thus possibly a plane of order 3 could be extended to a non-desarguean plane of order 12 or more.

6   Conclusion

Of course, if a non-desarguean projective plane of a given order does not exist, then no amount of subtle programming will matter. Nonetheless, it seems prudent to complement attempts to prove that they don't exist with searches to find them, in hopes that one or the other will succeed!

References

1. L. M. Batten, Combinatorics of Finite Geometries, 2nd edn., Cambridge University Press, Cambridge, 1997. 82
2. R. H. Bruck, *Finite nets II: uniqueness and embedding*, Pacific J. Math. **13**, (1963), 421–457. 84, 85
3. R. H. Bruck and H. J. Ryser, *The nonexistence of certain finite projective planes*, Canad. J. Math. **1**, (1949), 88–93. 83
4. G. Conant and A. Kruckman, *Independence in generic incidence structures*, arXiv:1709.09626 (2017). 84
5. N. G. de Bruijn and P. Erdös, *On a combinatorial problem*, Indag. Math. **10** (1948), 421–423. 84
6. P. Dembowski, Finite Geometries, Springer-Verlag, New York, 1968. 82
7. J. Dénes and A. D. Keedwell, Latin Squares and their Applications, Academic Press, New York, 1974. 84, 85
8. J. Dénes and A. D. Keedwell, Latin Squares: New Developments in the Theory and Applications, Annals of Discrete Math, vol. **46**, North-Holland, Amsterdam, 1991. 84, 85
9. S. Dow, *An improved bound for extending partial projective planes*, Discrete Math. **45**, (1943), 199–207. 84
10. R. Freese, *Lectures on projective planes*, http://math.hawaii.edu/~ralph. 83, 86
11. M. Hall, Jr., *Projective planes*, Trans. Amer. Math. Soc. **54**, (1943), 229–277. 82, 83
12. M. Hall, Jr., The Theory of Groups, Macmillan, New York, 1959. 82
13. C. W. H. Lam, L. Thiel and S. Swiercz, *The non-existence of finite projective planes of order* 10, Canad. J. Math. **41** (1989), 1117–1123. 83
14. J. B. Nation and J. Seffrood, *Dual linear spaces generated by a non-Desarguesian configuration*, Contributions to Discrete Math. **6** (2011), 98–141. 84, 85
15. S. S. Shrikhande, *A note on mutually orthogonal latin squares*, Sankhyā Ser. A **23** (1961), 115–116. 85

The role of twisted wreath products in the Finite Congruence Lattice Problem

Péter P. Pálfy

Alfréd Rényi Institute of Mathematics
Hungarian Academy of Sciences, Budapest, Hungary
ppp@renyi.hu

Abstract. The problem whether every finite lattice is representable as the congruence lattice of a finite algebra has been reduced to a group theoretic question: whether every finite lattice occurs as an interval in the subgroup lattice of a finite group. Based on works of R. Baddeley, A. Lucchini, F. Börner, J. Shareshian, and M. Aschbacher the problem can be further reduced to two particular cases: intervals in subgroup lattices of finite groups where the group is either almost simple or a twisted wreath product of a restricted type. So the group theoretic construction of twisted wreath products introduced by B. H. Neumann in 1963 seems to play a crucial role in dealing with the finite congruence lattice problem.

1 The Finite Congruence Lattice Problem

A famous unsolved problem in universal algebra asks whether every finite lattice is isomorphic to the congruence lattice of a finite algebra. Since finite lattices are obviously algebraic, it follows from the fundamental Grätzer–Schmidt Theorem [7] that every finite lattice is the congruence lattice of some algebra. However, all known proofs of the Grätzer–Schmidt Theorem construct infinite algebras in almost all cases. P. Pudlák and the author [13] have shown that the finiteness problem is equivalent to a group theoretic one:

Problem 1. Is every finite lattice isomorphic to an interval in the subgroup lattice of a finite group?

For a group G and a subgroup $H < G$ we write

$$\mathrm{Int}(H, G) = \{\, X \mid H \leq X \leq G \,\}$$

for the lattice of intermediate subgroups (in other words: overgroups of H), and call it the **interval** between H and G in the subgroup lattice.

One direction of the equivalence is obvious. Let G act on the set of right cosets of the subgroup H, and consider each permutation in G as an operation with one variable. Then the congruences are exactly the partitions into cosets of subgroups belonging to the interval $\text{Int}(H, G)$, hence the congruence lattice of this multi-unary algebra is isomorphic to this interval. Concerning the reverse implication, it should be emphasized that we do not claim that the congruence lattices of finite algebras are (up to isomorphism) the same as the intervals in subgroup lattices of finite groups. What we proved is that if *all* finite lattices can be represented as congruence lattices of finite algebras then *all* finite lattices can be represented as intervals in subgroup lattices of finite groups. In fact, we embed any finite lattice into a finite lattice with some useful properties, and then we show that the smallest algebra with a congruence lattice having these properties is a transitive permutation group considered as a multi-unary algebra.

It was shown by Jiří Tůma [17] that every algebraic lattice is isomorphic to an interval in the subgroup lattice of an infinite group. So it is the finiteness of the group what seems to constitute a severe restriction. Therefore, it is generally believed that the answer to the finite congruence lattice problem is negative.

2 Twisted Wreath Products

The notion of twisted wreath product was introduced by B. H. Neumann [10] in 1963. At first glance his definition looks quite complicated. M. Suzuki [16, Chapter 2, §10] presented a more elegant treatment of this construction. In [12] we gave a natural explanation for the occurrence of twisted wreath products. Although originally Neumann used twisted wreath products for constructing infinite groups with peculiar properties, here in the present paper we will stick to finite groups.

Twisted wreath products occur in the O'Nan–Scott–Aschbacher Theorem on the classification of primitive finite permutation groups. They were erroneously omitted from the first version [14] of the theorem, and were only added later to the list in the paper of Michael Aschbacher and Leonard Scott [2], and independently by László Kovács [8]. (See also [9].)

The fundamental role of twisted wreath products in the problem of representing finite lattices as intervals in subgroup lattices of finite groups was explicitly or implicitly observed in the papers of Robert Baddeley and Andrea Lucchini [4], Baddeley [3], Ferdinand Börner [5], John Shareshian [15], and Michael Aschbacher [1].

The ingredients of the twisted wreath product are the following: (finite) groups D (the domain) and T (the target), a subgroup $D_0 \leq D$ and a homomorphism $\varphi : D_0 \to \mathrm{Aut}(T)$ into the automorphism group of T. Let us decompose $D = D_0 x_1 \cup D_0 x_2 \cup \cdots \cup D_0 x_m$ into a disjoint union of right cosets. Now let

$$\mathrm{Sdp}(D_0, \varphi) = \{f : D \to T \mid f(ax_i) = \varphi_a(t_i),\ a \in D_0,\ t_i \in T,\ 1 \leq i \leq m\}.$$

It is easy to check that $\mathrm{Sdp}(D_0, \varphi)$ is a D-invariant subdirect product in T^D, where D acts on T^D via the natural action $f^d(x) = f(xd^{-1})$ for $f \in T^D$, $d, x \in D$. The **twisted wreath product** of T and D with respect to the subgroup $D_0 \leq D$ and the homomorphism $\varphi : D_0 \to \mathrm{Aut}(T)$ is defined as the semidirect product

$$\mathrm{Twr}(T, D, D_0, \varphi) = \mathrm{Sdp}(D_0, \varphi) \rtimes D.$$

3   The Reduction Theorem

Slighly improving Börner's result [5, Theorem 6.1] — by using a different lattice embedding lemma — we gave a proof [12] of the following reduction theorem.

Theorem 1. *Every finite lattice is isomorphic to an interval in the subgroup lattice of a finite group if and only if one of the following is true:*

(1) Every finite lattice consisting of more than one element is isomorphic to an interval $\mathrm{Int}(H, G)$ *in the subgroup lattice of an* **almost simple** *finite group* G *with a core-free subgroup* H *(that is,* $\bigcap_{g \in G} g^{-1} H g = 1$*).*

(2) Every finite lattice consisting of more than one element is isomorphic to an interval $\mathrm{Int}(D, G)$ *in the subgroup lattice of a* **twisted wreath product** $G = \mathrm{Twr}(T, D, D_0, \varphi)$ *of a non-abelian finite simple group* T *and a finite group* D *with respect to a subgroup* $D_0 < D$ *and a homomorphism* $\varphi : D_0 \to \mathrm{Aut}(T)$ *satisfying* $\varphi(D_0) \geq \mathrm{Inn}(T)$*, the group of inner automorphisms of* T*.*

With some extra work one can show also (as it was done by Börner [5]) that in case (2) we can force D_0 to be core-free in D.

We should note that the proof uses the classification of finite simple groups via one of its well-known consequences, **Schreier's Hypothesis**, claiming that the outer automorphism group $\mathrm{Out}(T) = \mathrm{Aut}(T) / \mathrm{Inn}(T)$ of every finite non-abelian simple group T is solvable.

As for many questions in finite group theory it would be desirable to reduce the problem to case (1) of **almost simple groups** (groups G

with a simple normal subgroup T with $\mathbf{C}_G(T) = 1$). However, it seems inevitable to consider also certain twisted wreath products.

On the lattice theoretical side the proof does not use any deep considerations. If there is a lattice L_1 not representable with an almost simple group as in case (1) of the theorem, and another lattice L_2 that cannot be represented as an interval as in case (2), then one constructs a lattice that cannot be represented as an interval in the subgroup lattice of any finite group, see Figure 1. (Here L^d denotes the dual of the lattice L and $\hat{L}$ refers to a suitable extension of L that is generated by its coatoms and contains L as a filter.)

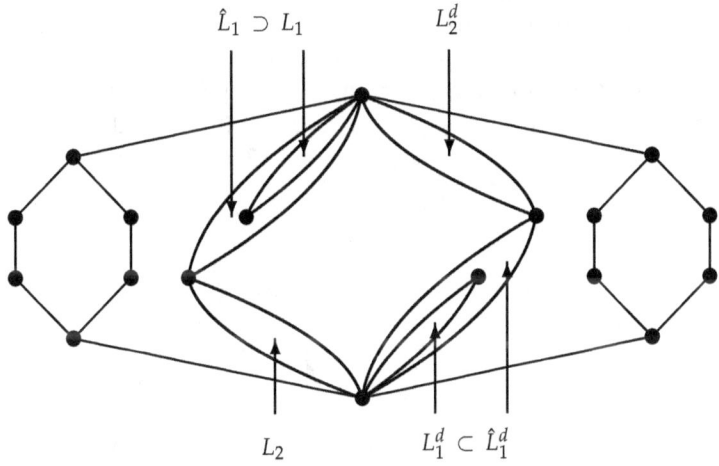

$$\hat{L}_1 \supset L_1 \qquad\qquad L_2^d$$

$$L_2 \qquad\qquad L_1^d \subset \hat{L}_1^d$$

Fig. 1. A possibly non-representable lattice

4   Intervals in the Subgroup Lattice of a Twisted Wreath Product

In case (2) of Theorem 1 one can describe the interval $\mathrm{Int}(D, G)$ in the following way. If $D < X \leq G = \mathrm{Twr}(T, D, D_0, \varphi)$, then $X = \mathrm{Sdp}(D_1, \varphi_1) \rtimes D$ for some subgroup $D_0 \leq D_1 \leq D$ and homomorphism $\varphi_1 : D_1 \to \mathrm{Aut}(T)$ extending φ. Moreover, $\mathrm{Sdp}(D_1, \varphi_1) \rtimes D \leq \mathrm{Sdp}(D_2, \varphi_2) \rtimes D$ iff $D_1 \geq D_2$ and $\varphi_1|_{D_2} = \varphi_2$. Hence we obtain:

Theorem 2. *Let* $G = \mathrm{Twr}(T, D, D_0, \varphi)$ *be the twisted wreath product of a non-abelian finite simple group* T *and a finite group* D *with respect to a subgroup* $D_0 < D$ *and a homomorphism* $\varphi : D_0 \to \mathrm{Aut}(T)$ *satisfying* $\varphi(D_0) \geq \mathrm{Inn}(T)$. *Then the interval* $\mathrm{Int}(D, G)$ *in the subgroup lattice of* G *is dually isomorphic to the lattice formed by all extensions of* φ *to subgroups of* D *together with a largest element added.*

The largest element on the top of all extensions corresponds to $D \in \mathrm{Int}(D, G)$ by the dual isomorphism.

For example, let A_5 and S_5 denote the alternating and the symmetric group of degree 5, and let $T = A_5$, $D = S_5 \times A_5$, $D_0 = \mathrm{diag}(A_5) = \{(a, a) \mid a \in A_5\} < D$, and fix an embedding $\varphi : D_0 \cong A_5 \to \mathrm{Aut}(T) \cong S_5$. It is easy to see that the subgroups of D containing D_0 are $D_0 = \mathrm{diag}(A_5)$, $A_5 \times A_5$, and $D = S_5 \times A_5$. Now φ has two extensions to $A_5 \times A_5$, corresponding to the first and the second projection. Likewise, there are two extensions to $S_5 \times A_5$. Together with the additional top element this gives a hexagon lattice, see Figure 2 (where $a, b \in A_5$, $s \in S_5$).

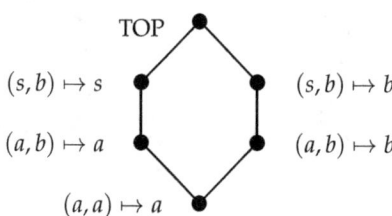

$$\begin{array}{c}\text{TOP} \\ (s,b) \mapsto s \qquad\qquad (s,b) \mapsto b \\ (a,b) \mapsto a \qquad\qquad (a,b) \mapsto b \\ (a,a) \mapsto a \end{array}$$

Fig. 2. A representation for the hexagon lattice

Hence by Theorem 2 the interval $\mathrm{Int}(D, G)$ in the subgroup lattice of $G = \mathrm{Twr}(A_5, S_5 \times A_5, \mathrm{diag}(A_5), \varphi)$ is the hexagon lattice.

Actually, Aschbacher was motivated by a paper of Yasuo Watatani, [18], where it was proved that whenever a lattice can be represented as an interval in a subgroup lattice of a finite group, then it also occurs as a lattice of intermediate subfactors of a von Neumann algebra. With the exception of two lattices, Watatani was able to find intervals isomorphic to every lattice with at most six elements. One of the missing cases was the hexagon lattice. Aschbacher [1] gave a general construction whose particular cases provided examples for the hexagon and for the other six-element lattice Watatani was not able to handle. Aschbacher's example was slighly different from ours, he used $D = A_6 \times A_6$ instead of $S_5 \times A_5$ (but the same T, D_0, and φ). The hexagon also occurs in the subgroup lattice of a simple group, for example, as the interval of overgroups of a solvable subgroup of order 55 in the alternating group A_{11}, see [11, p. 477].[1]

5 Open Cases

William DeMeo [6] found representations of all lattices consisting of at most 7 elements, with two exceptions shown in Figure 3.

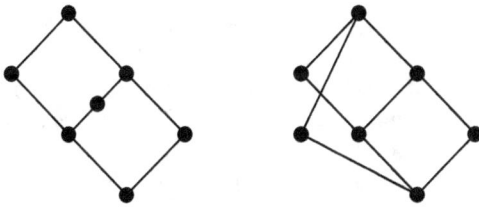

Fig. 3. Open cases

So currently these are the smallest lattices for which no representation as an interval in the subgroup lattice of a finite group is known. (However, he showed that the lattice on the left hand side is the congruence lattice of a finite algebra.)

John Shareshian [15] suggested some candidates for lattices that may not be representable as intervals in subgroup lattices of finite groups. The smallest among these lattices is shown in Figure 4.

Acknowledgement. The author has been supported by the National Research, Development and Innovation Fund of Hungary, grant no. 115799.

References

1. M. Aschbacher, On intervals in subgroup lattices of finite groups, *Journal of the American Mathematical Society* **21** (2008), 809–830. 89, 92

[1] I am very grateful to the referee for calling my attention to this example from my own old paper that I had forgotten.

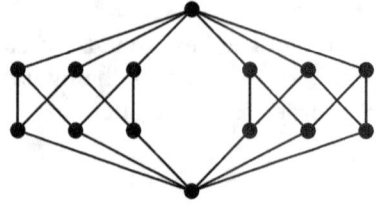

Fig. 4. A lattice conjectured not to be representable

2. M. Aschbacher and L. L. Scott, Maximal subgroups of finite groups, *Journal of Algebra* **92** (1985), 44–80. 89

3. R. Baddeley, A new approach to the finite lattice representation problem, *Periodica Mathematica Hungarica* **36** (1998), 17–59. 89

4. R. Baddeley and A. Lucchini, On representing finite lattices as intervals in subgroup lattices of finite groups, *Journal of Algebra* **196** (1997), 1–100. 89

5. F. Börner, A remark on the finite lattice representation problem, in *Contributions to General Algebra, 11* (I. Chajda et al., eds.) (Verlag Johannes Heyn, Klagenfurt, 1999), 5–38. 89, 90

6. W. DeMeo, *Congruence lattices of finite algebras*, (PhD Thesis, University of Hawai'i, 2012) arXiv:1204.4305 93

7. G. Grätzer and E. T. Schmidt, Characterizations of congruence lattices of abstract algebras, *Acta Scientiarum Mathematicarum (Szeged)* **24** (1963), 34–59. 88

8. L. G. Kovács, Maximal subgroups in composite finite groups, *Journal of Algebra* **99** (1986), 114–131. 89

9. M. W. Liebeck, C. E. Praeger and J. Saxl, On the O'Nan–Scott theorem for finite primitive permutation groups, *Journal of the Australian Mathematical Society, Ser. A* **44** (1988), 389–396. 89

10. B. H. Neumann, Twisted wreath product of groups, *Archiv der Mathematik* **14** (1963), 1–6. 89

11. P. P. Pálfy, On Feit's examples of intervals in subgroup lattices, *Journal of Algebra* **116** (1988), 471–480. 93

12. P. P. Pálfy, Subgroups of twisted wreath products, to appear in the Proceedings of the conference *Groups St Andrews 2017 in Birmingham*. 89, 90

13. P. P. Pálfy and P. Pudlák, Congruence lattices of finite algebras and intervals in subgroup lattices of finite groups, *Algebra Universalis* **11** (1980), 22–27. 88

14. L. L. Scott, Representations in characteristic *p*, in *Santa Cruz Conference on Finite Groups*, Proc. Sympos. Pure Math. **37** (American Mathematical Society, Providence, R.I., 1980), 318–331. 89

15. J. Shareshian, Topology of order complexes of intervals in subgroup lattices, *Journal of Algebra* **268** (2003), 677–686. 89, 93

16. M. Suzuki, *Group Theory I*, Grundlehren der mathematischen Wissenschaften **247** (Springer-Verlag, Berlin–Heidelberg–New York, 1982). 89

17. J. Tůma, Intervals in subgroup lattices of infinite groups, *Journal of Algebra* **125** (1989), 367–399. 89
18. Y. Watatani, Lattices of intermediate subfactors, *Journal of Functional Analysis* **140** (1996), 312–334. 92

Ω-algebras

Branimir Šešelja* and Andreja Tepavčević

[1] Faculty of Sciences, University of Novi Sad, Serbia
seselja@dmi.uns.ac.rs
[2] Mathematical Institute of the Serbian Academy of Sciences and Arts, Belgrade
andreja@dmi.uns.ac.rs

Abstract. Starting with Ω-sets where Ω is a complete lattice, we introduce the notion of an Ω-algebra. This is a classical algebra equipped with an Ω-valued equality replacing the ordinary one. In these new structures identities hold as appropriate lattice-theoretic formulas. Our investigation is related to weak congruences of the basic algebra to which a generalized equality is associated. Namely every Ω-algebra uniquely determines a closure system in the lattice of weak congruences of the basic algebra. By this correspondence we formulate a representation theorem for Ω-algebras.

Keywords: Ω-algebra, Ω-group, lattice-valued, weak congruence.

1 Introduction

The topic of this research is Ω-valued algebraic structures, where Ω is a complete lattice.

Our research originates in the theory of Ω-sets. These structures appeared in 1979 in the paper [6] by Fourman and Scott. Introducing Ω-sets, they intended to use them for modeling intuitionistic logic, analogously to the application of Boolean-valued models in first-order logic. An Ω-set is a nonempty set A equipped with an Ω-valued equality E, with truth-values in a complete Heyting algebra Ω. E is a symmetric and transitive function from A^2 to Ω. In this framework, Ω-sets consist of so called 'partial elements', since $E(a, a)$ is understood as a 'probability' of $a \in A$, and E is not reflexive (not constantly equal 1 for pairs (x, x)). Ω-sets have been further applied to non-classical predicate logics, and also partially in theoretical foundations of fuzzy set theory ([7,8]).

---

*Research supported by Ministry of Education and Science, Republic of Serbia, Grant No. 174013.

Another source of our investigation is the concept of algebras with fuzzy equality, introduced by Bělohlávek and Vychodil ([1]). Following the philosophy of fuzzy mathematics, they use a complete residuated lattice L as a truth-values structure (called also a membership-values structure) and equip a nonempty set A with a particular L-valued equality which should replace the characteristic function of the classical equality. Adding operations to this structure they obtain so-called L-algebras. The corresponding equational logic is the one by Pavelka ([10]). Basic parts of universal algebra are presented in this framework, including a Birkoff-like variety theorem. Observe that here an L-valued equality is reflexive (equal 1 for pairs (x, x), $x \in A$).

Algebraic topics have been investigated in the framework of lattice-valued structures, see e.g., Kuraoka and Suzuki [9]. A generalized equality was used in particular by Demirci ([5]), Bělohlávek and Vychodil ([1]) and others.

Introducing Ω-algebras, we use Ω-sets and in our approach Ω is a complete lattice (not necessarily a Heyting algebra). A reason for this membership-values structure is that it allows the use of cut-sets as a tool appearing in the fuzzy set theory. In this setting, main algebraic and set-theoretic notions and their properties can be generalized from their classical origin to the lattice-valued framework. So we deal also with lattice-valued structures. Still the main reason for using a complete lattice as a co-domain comes from the representation theorems that we prove here. In this construction the lattice of truth-values for an Ω-algebra is closely related to the weak-congruence lattice of the basic, underlying algebra and it could be any algebraic lattice.

Identities for lattice-valued structures with a fuzzy equality were introduced in [1] with graded satisfiability. Our approach was introduced in [13], and then developed in [2]. In this framework, an identity holds if the corresponding lattice-theoretic formula is fulfilled. An identity may hold on a lattice-valued algebra, while the underlying classical algebra need not satisfy the same identity.

In the present note we give basic features for Ω-algebras and we prove representation theorems for Ω-algebras in general.

2 Preliminaries

2.1 Notation

An **algebra** is denoted by $\mathcal{A} = (A, F)$, where A is a nonempty set and F is a set of (fundamental) operations on A. We deal with **terms, term-operations**, and **identities** in the given language as formulas $t_1 \approx t_2$, where t_1, t_2 are terms in the same language.

In addition to **congruences**, we use **weak congruences** on $\mathcal{A}$ as symmetric and transitive subalgebras of $\mathcal{A}^2$; a weak congruence on $\mathcal{A}$ is obviously a congruence on the subalgebra determined by its domain. The collection $\mathrm{Con_w}(\mathcal{A})$ of all weak congruences on an algebra $\mathcal{A}$ is an algebraic lattice under inclusion ([11,3,4]).

2.2 Ω-valued sets and relations

By $(\Omega, \wedge, \vee, \leqslant)$ we denote a complete lattice with the top and the bottom elements, 1 and 0, respectively.

If A is a nonempty set, then an **Ω-valued function** μ on A is a map $\mu : A \to \Omega$. For $x \in A$, $\mu(x)$ is a **degree of membership** of x to μ.

For $p \in L$, a **cut set** or a *p*-**cut** of an Ω-valued function $\mu : A \to \Omega$ is a subset μ_p of A which is the inverse image of the principal filter $\uparrow p$ in Ω: $\mu_p = \mu^{-1}(\uparrow p) = \{x \in X \mid \mu(x) \geqslant p\}$.

An **Ω-valued** (binary) **relation** R on A is an Ω-valued function on A^2, i.e., it is a mapping $R : A^2 \to \Omega$. As above, for $p \in \Omega$, a cut R_p of R is the binary relation on A, which is the inverse image of $\uparrow p$: $R_p = R^{-1}(\uparrow p)$. R is **symmetric** if $R(x,y) = R(y,x)$ for all $x, y \in A$, and **transitive** if $R(x,y) \geqslant R(x,z) \wedge R(z,y)$ for all $x, y, z \in A$.

Lemma 1 ([12]). *An Ω-valued binary relation R on A is symmetric (transitive) if and only if all cuts of R are classical symmetric (transitive) relations on A.*

A symmetric and transitive Ω-valued relation on A fulfills the **strictness** property:

$$R(x,y) \leqslant R(x,x) \wedge R(y,y), \tag{1}$$

Strictness can be understood as a *weak reflexivity* of R. Therefore, a symmetric and transitive Ω-valued relation on A is a **weak Ω-valued equivalence on A**.

If $\mu : A \to \Omega$ is an Ω-valued function on A, then the map $R : A^2 \to \Omega$ on A is an **Ω-valued relation on** μ if for all $x, y \in A$

$$R(x,y) \leqslant \mu(x) \wedge \mu(y). \tag{2}$$

An ordinary symmetric and transitive relation is reflexive on its domain. Analogously, an Ω-valued relation R on $\mu : A \to \Omega$ is said to be **reflexive on** μ if

$$R(x,x) = \mu(x) \text{ for every } x \in A. \tag{3}$$

A symmetric and transitive Ω-valued relation R on A, which is reflexive on $\mu : A \to \Omega$ is an Ω-**valued equivalence on** μ.

By (1), if $R : A^2 \to \Omega$ is a weak Ω-valued equivalence on A, then it is an Ω-valued equivalence on $\mu : A \to \Omega$, such that $\mu(x) = R(x,x)$. The Ω-valued function μ is said to be **determined** by R.

A weak Ω-valued equivalence R on A is a weak Ω-**valued equality**, if it satisfies the **separation property**:

$$\text{If } R(x,x) \neq 0, \text{ then } R(x,y) = R(x,x) \text{ implies } x = y. \tag{4}$$

Remark 1. The *separation* property is in [6] introduced by a weaker condition:

$$R(x,y) = R(x,x) = R(y,y) \text{ implies } x = y.$$

Analogously, an Ω-valued equivalence on $\mu : A \to \Omega$ satisfying (4) is an Ω-**valued equality** on μ.

If $A = (A, F)$ is an algebra and $\mu : A \to \Omega$ an Ω-valued function on A, then μ is **compatible** with the operations in F, if for every n-ary operation $f \in F$, for all $a_1, \ldots, a_n \in A$, and for every constant (nullary operation) $c \in F$

$$\bigwedge_{i=1}^{n} \mu(a_i) \leqslant \mu(f(a_1, \ldots, a_n)), \text{ and } \mu(c) = 1. \tag{5}$$

Further, an Ω-valued relation $R : A^2 \to \Omega$ on A is **compatible** with the operations in F if for every n-ary operation $f \in F$, for all $a_1, \ldots, a_n,$ $b_1, \ldots, b_n \in A$, and for every constant $c \in F$

$$\bigwedge_{i=1}^{n} R(a_i, b_i) \leqslant R(f(a_1, \ldots, a_n), f(b_1, \ldots, b_n)), \text{ and } R(c,c) = 1. \tag{6}$$

The following is straightforward.

Lemma 2. *Let $\mathcal{A} = (A, F)$ be an algebra.*

An Ω-valued function $\mu : A \to \Omega$ on A is compatible with all the operations in F, if and only if for every $p \in \Omega$, μ_p is a subalgebra of $\mathcal{A}$.

Similarly, an Ω-valued relation $R : A^2 \to \Omega$ on A is compatible with all the operations in F, if and only if for every $p \in \Omega$, R_p is compatible with all the operations in F.

3   Ω-algebras

3.1   Ω-set and Ω-algebra; identities

An Ω-**set** (originating from [6]) is a pair (A, E), where A is a nonempty set, and E is a symmetric and transitive Ω-valued relation on A, fulfilling the separation property (4).

We also say that (A, E) is a **lattice-valued set**, without particularly fixing the co-domain lattice. As defined above, the Ω-valued function $\mu : A \to \Omega$ on A, given by $\mu(x) = E(x, x)$, is determined by E, which is a weak Ω-valued equivalence on A. But E is also an Ω-valued equality on μ. Therefore, we say that in an Ω-set (A, E), E is an Ω-**valued equality**.

Lemma 3. *Every cut E_p, $p \in \Omega$, of the Ω-valued equality in an Ω-set (A, E) is an equivalence relation on the corresponding cut μ_p of μ.*

A pair $\overline{\mathcal{A}} = (\mathcal{A}, E)$ is an Ω-**algebra** if $\mathcal{A} = (A, F)$ is an algebra, (A, E) is an Ω-set and E is compatible with the operations in F. $\mathcal{A}$ is the **underlying, basic algebra** of $\overline{\mathcal{A}}$.

If we do not fix Ω, we say that $\overline{\mathcal{A}} = (\mathcal{A}, E)$ is a **lattice-valued algebra**.

Proposition 1 ([2]). *Let $(\mathcal{A}, E)$ be an Ω-algebra. Then:*

(i) The Ω-valued function μ determined by E is compatible with the fundamental operations on $\mathcal{A}$.

(ii) For every $p \in \Omega$, the cut μ_p of μ is a subalgebra of $\mathcal{A}$, and

(iii) Every cut of E is a weak congruence on $\mathcal{A}$, namely for $p \in E$, E_p is a congruence on μ_p.

Identities hold on Ω-algebras in a particular way, as introduced in [13]. Recall that in the equational logic, the relational symbol $\approx$ in an identity $u \approx v$ is modeled by the classical equality " $=$ ". In the framework of Ω-algebras, this relational symbol corresponds to the Ω-equality E, as follows.

Let $(\mathcal{A}, E)$ be an Ω-algebra and $u(x_1, \ldots, x_n) \approx v(x_1, \ldots, x_n)$, briefly $u \approx v$ be an identity in the type of $\mathcal{A}$. We assume, as usual, that variables appearing in terms u and v are from $x_1, \ldots, x_n$. Then, $(\mathcal{A}, E)$ **satisfies identity** $u \approx v$ (i.e., this identity **holds** on $(\mathcal{A}, E)$) if

$$\bigwedge_{i=1}^{n} \mu(a_i) \leqslant E(u(a_1, \ldots, a_n), v(a_1, \ldots, a_n)), \tag{7}$$

for all $a_1, \ldots, a_n \in A$ and the term-operations on $\mathcal{A}$ corresponding to terms u and v respectively.

If Ω-algebra $(\mathcal{A}, E)$ satisfies an identity, this identity need not hold on $\mathcal{A}$, but the converse holds: *An identity $u \approx v$ fulfilled on an algebra $\mathcal{A}$ holds on an Ω-algebra $(\mathcal{A}, E)$ as well.*

Theorem 1 ([2]). *Let $(\mathcal{A}, E)$ be an Ω-algebra, and $\mathcal{F}$ a set of identities in the language of $\mathcal{A}$. Then, $(\mathcal{A}, E)$ satisfies all identities in $\mathcal{F}$ if and only if for every $p \in L$ the quotient algebra μ_p / E_p satisfies the same identities.*

Corollary 1. *If a diagonal relation $\Delta_A = \{(a, a) \mid a \in A\}$ is a cut of E, then each identity fulfilled by an Ω-algebra $\overline{\mathcal{A}} = (\mathcal{A}, E)$ also holds on the underlying algebra $\mathcal{A}$.*

Proof. Follows by Theorem 1: if $E_p = \Delta_A$ for some $p \in \Omega$, then quotient algebra μ_p / E_p is isomorphic to $\mathcal{A}$.   □

By Corollary 1, we are interested in Ω-algebras which do not contain a copy of the underlying algebra among quotient substructures. An Ω-algebra $\overline{\mathcal{A}} = (\mathcal{A}, E)$ is said to be **proper** if Δ_A is not a cut of E.

Theorem 2. $\overline{\mathcal{A}} = (\mathcal{A}, E)$ *is a proper Ω-algebra if and only if*

$$\text{there are } a, b \in A, a \neq b, \text{ such that } E(a, b) \geqslant \bigwedge \{E(x, x) \mid x \in A\}. \quad (8)$$

Proof. Suppose that for all a, b, $a \neq b$, we have $E(a, b) \notin \uparrow \bigwedge \{E(x, x) \mid x \in A\}$. Then for $p = \uparrow \bigwedge \{E(x, x) \mid x \in A\}$, the cut E_p does not contain any pair (a, b) with $a \neq b$. Obviously, for every $z \in A$, $(z, z) \in E_p$. Hence, $\overline{\mathcal{A}}$ is not a proper Ω-algebra.

Conversely, suppose that $\overline{\mathcal{A}}$ is not a proper Ω-algebra, i.e., that Δ_A is a cut of E, $\Delta_A = E_p$ for some $p \in \Omega$. Then $E(x, x) \geqslant p$ for every $x \in A$ and $\bigwedge_{x \in A} E(x, x) \geqslant p$. Since $\Delta_A = E_p$, there are no a, b, $a \neq b$, such that $(a, b) \in E_p$. Therefore, $E(a, b) \not\geqslant p$ and (8) does not hold.   □

3.2 Representation

Recall that a **closure system** on a nonempty set X is a collection of subsets of X closed under set intersections (including $\bigcap \emptyset$ which is X). It is a complete lattice under inclusion.

Proposition 2. *The collection of cuts of E in an Ω-algebra $\overline{\mathcal{A}} = (\mathcal{A}, E)$ is a closure system on A^2, a subposet of the weak congruence lattice $\mathrm{Con_w}(\mathcal{A})$ of $\mathcal{A}$.*

Proof. By (*iii*) in Proposition 1, the cuts of E are weak congruences on $\mathcal{A}$. The collection of cuts is closed under intersection, namely, $\bigcap_i E_{p_i} = E_{\bigvee_i p_i}$, which is a property of every lattice-valued function (see e.g., [12]). The greatest element is the cut $E_0 = A^2$.   □

Theorem 3 (Representation). *Let $\mathcal{A}$ be an algebra and $\mathcal{R}$ a closure system in $\mathrm{Con_w}(\mathcal{A})$ such that*

$$\text{if } a \neq b, \text{ then } (a,b) \notin \bigcap \{R \in \mathcal{R} \mid (a,a) \in R\} \quad \text{for all } a, b \in A. \quad (9)$$

Then there is a complete lattice Ω and an Ω-algebra $(\mathcal{A}, E)$ with the underlying algebra $\mathcal{A}$, such that $\mathcal{R}$ consists of cuts of E.

Proof. We take Ω to be the starting collection $\mathcal{R}$ of weak congruences ordered by the dual of inclusion, $\supseteq$. Being a closure system, $(\mathcal{R}, \supseteq)$ is a complete lattice. Next, we define $E : A^2 \to \Omega$:

$$E(a,b) := \bigcap (R \in \mathcal{R} \mid (a,b) \in R) \quad \text{for all } a, b \in A. \quad (10)$$

Now we have that $E_R = R$ (the cut determined by R considered as an element of Ω, coincides with R as a weak congruence), which could be checked straightforwardly. Next, by Lemma 1 and Lemma 2, E is symmetric, transitive Ω-valued relation on A, compatible with the operations in F. Now we prove the separation property, i.e., that $E(x,x) \neq 0$ and $x \neq y$ imply that $E(x,y) < E(x,x)$. By strictness we have $E(x,y) \leqslant E(x,x)$. Now suppose there are $x, y \in A$, such that $x \neq y$ and $E(x,y) = E(x,x)$. Then

$$E(x,y) = \bigcap (R \in \mathcal{R} \mid (x,y) \in R) = \bigcap (R \in \mathcal{R} \mid (x,x) \in R).$$

Hence $(x,y) \in \bigcap(R \in \mathcal{R} \mid (x,x) \in R)$, which contradicts (9).   □

For a symmetric and transitive relation $R \subseteq A^2$, we denote by $\mathrm{dom}R$ the set $\{x \in A \mid (x,x) \in R\}$.

Corollary 2. *Let $\mathcal{A}$ be an algebra and $\mathcal{R}$ a closure system in $\mathrm{Con_w}(\mathcal{A})$ fulfilling condition (9). Let also $\mathcal{F}$ be a set of identities in the language of $\mathcal{A}$ and suppose that for every $R \in \mathcal{R}$, the algebra $\mathrm{dom}R/R$ fulfills these identities. Then there is a complete lattice Ω and an Ω-algebra $(\mathcal{A}, E)$, such that $\mathcal{R}$ consists of cuts of E and $(\mathcal{A}, E)$ satisfies $\mathcal{F}$.*

Proof. As in Theorem 3, $\Omega = (\mathcal{R}, \supseteq)$, and $E : A^2 \to \Omega$ is defined by (10). For cuts of E we have $E_R = R$, with $R \in \mathcal{R}$. Next, we define $\mu : A \to \Omega$ by $\mu(x) = E(x,x)$, and we get $\mathrm{dom}R = \mu_R$. Therefore, for every $R \in \mathcal{R}$, we have $\mathrm{dom}R/R = \mu_R/E_R$. Since by assumption all these quotient algebras fulfill identities in $\mathcal{F}$, then by Theorem 1, the constructed Ω-algebra $(\mathcal{A}, E)$ also satisfies these identities.   □

The lattice Ω and the corresponding Ω-algebra described in the proof of Theorem 3 are said to be obtained by the **canonical construction** over the algebra $\mathcal{A}$.

Algebraic properties of Ω-algebras are related to their quotient structures over cuts of the Ω-valued equality E. Suppose that we have different complete lattices, Ω_1 and Ω_2 and an algebra $\mathcal{A}$. Let $(\mathcal{A}, E1)$ and $(\mathcal{A}, E2)$ be an Ω_1-valued algebra and an Ω_2-valued algebra respectively. We say that the structures $(\mathcal{A}, E1)$ and $(\mathcal{A}, E2)$ are **cut-equivalent** if their collections of quotient algebras over cuts of $E1$ and $E2$ coincide, i.e., if for every $p \in \Omega_1$ there is $q \in \Omega_2$ such that $\mu 1_p / E1_p = \mu 2_q / E2_q$ and vice versa.

We prove in the sequel that for every Ω-algebra there is a cut-equivalent one obtained by the canonical construction. To do this, we need the following lemma.

Lemma 4. *Let $\Omega 1$ and $\Omega 2$ be complete lattices, $\mathcal{A} = (A, F)$ an algebra and $(\mathcal{A}, E1)$ and $(\mathcal{A}, E2)$ $\Omega 1$-algebra and $\Omega 2$-algebra, respectively. If the collection of cuts of $E1$ and $E2$ coincide, then $(\mathcal{A}, E1)$ and $(\mathcal{A}, E2)$ are cut-equivalent.*

Proof. Clearly, if the cuts of $E1$ and $E2$ coincide, then also the corresponding domains of these weak congruences coincide (since $\mu(x) = E(x, x)$), and therefore the quotient algebras are also the same. □

Theorem 4. *Let $\overline{\mathcal{A}} = (\mathcal{A}, E)$ be an Ω-algebra where Ω is an arbitrary complete lattice. Then there is a lattice and a lattice-valued algebra cut-equivalent with $\overline{\mathcal{A}}$, obtained by the canonical construction over $\mathcal{A}$.*

Proof. Let Ω be a fixed complete lattice and $(\mathcal{A}, F)$ be an Ω-algebra, with the collection $\{E_p \mid p \in \Omega\}$ of cuts of E. We define a new lattice $\Omega 1$ by $\Omega 1 = (\{E_p \mid p \in \Omega\}, \supseteq)$. This lattice is complete since the collection of cuts is a closure system. Next we define $E1 : A^2 \to \Omega 1$ by $E1(x, y) := \bigcap(R \in \Omega 1 \mid (x, y) \in R)$. $E1$ is clearly well defined. In addition, the cuts of $E1$ coincide with the cuts of E since by the construction, if $E_p \in \Omega 1$, for the corresponding cut of $E1$ we have $E1_{E_p} = E_p$. We use the same construction of $E1$ as in Theorem 3, therefore $E1$ is a separated $\Omega 1$-equality. So, $(\mathcal{A}, E1)$ is an $\Omega 1$-algebra obtained by the canonical construction over $\mathcal{A}$ and by Lemma 4, $(\mathcal{A}, E)$ and $(\mathcal{A}, E1)$ are cut-equivalent. □

Example 1. The lattice Ω is given in Fig.1., and the five-element algebra is $\mathcal{A} = (\{e, a, b, c, d\}, \cdot, ', e)$ with a constant e, a binary $(\cdot)$ and a unary $(')$ operation given by the tables. The Ω-algebra is $(\mathcal{A}, E)$, the table of E is also provided, together with its diagonal, compatible function μ.

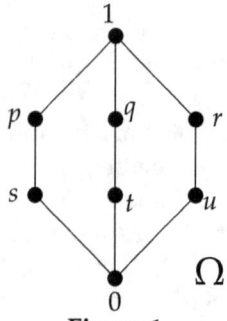

Figure 1

·	e	a	b	c	d
e	e	a	b	c	d
a	a	e	d	a	a
b	b	c	c	e	a
c	c	c	e	b	b
d	d	d	d	d	e

′	e	a	b	c	d
	e	a	c	b	d

E	e	a	b	c	d
e	1	u	t	t	s
a	u	r	0	0	0
b	t	0	q	t	0
c	t	0	t	q	0
d	s	0	0	0	p.

$$\mu = \begin{pmatrix} e & a & b & c & d \\ 1 & r & q & q & p \end{pmatrix}.$$

The cuts of E are either diagonal relations on subalgebras (E_q on $\{e, b, c\}$ and E_r on $\{e, a\}$), or they are full relations on one-, two- or three-element subalgebras (e.g., E_t is a full relation on $\{e, b, c\}$). Trivially, E_0 is a full relation on the whole algebra. All the corresponding quotient algebras are groups, hence $(\mathcal{A}, E)$ is an Ω-group. Observe that the basic five-element algebra is not a group.

Example 2. Let Ω be the lattice given by the diagram in Fig.2.

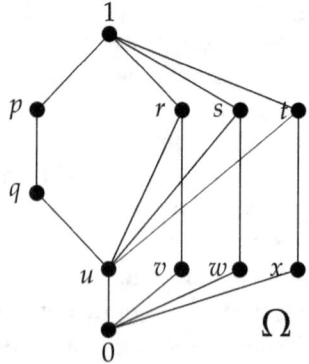

Figure 2

Consider the symmetric group S_3 (given by the table) as the underlying algebra. The corresponding Ω-group is (S_3, E) where E is given in the sequel. (S_3, E) is a *commutative* Ω-group, while the basic, underlying algebra is not commutative.

∘	e	f	g	h	j	k
e	e	f	g	h	j	k
f	f	e	h	g	k	j
g	g	j	e	k	f	h
h	h	k	f	j	e	g
j	j	g	k	e	h	f
k	k	h	j	f	g	e

E	e	f	g	h	j	k
e	1	x	w	q	q	v
f	x	t	u	0	0	u
g	w	u	s	0	0	u
h	q	0	0	p	q	0
j	q	0	0	q	p	0
k	v	u	u	0	0	r

$$\mu = \begin{pmatrix} e & f & g & h & j & k \\ 1 & t & s & p & p & r \end{pmatrix}.$$

All the structures μ_z / E_z, $z \in \Omega$ are groups of order 3, 2 or 1, hence Abelian. Therefore, this structure is an Abelian Ω-group, identity $x \cdot y \approx y \cdot x$ holds as the formula $\mu(x) \wedge \mu(y) \leqslant E(x \cdot y, y \cdot x)$.

For the cuts, we have e.g., $\mu_p = \{e, h, j\}$, $\mu_u = \{e, f, g, h, j, k\}$.

E_p	e	f	g	h	j	k
e	1	0	0	0	0	0
f	0	0	0	0	0	0
g	0	0	0	0	0	0
h	0	0	0	1	0	0
j	0	0	0	0	1	0
k	0	0	0	0	0	0

E_u	e	f	g	h	j	k
e	1	0	0	1	1	0
f	0	1	1	0	0	1
g	0	1	1	0	0	1
h	1	0	0	1	1	0
j	1	0	0	1	1	0
k	0	1	1	0	0	1

Hence, E_p is a weak congruence on S_3, a diagonal of $\mu_p = \{e, h, j\}$ and μ_p / E_p is a group of order 3. Next, μ_u is the underlying group S_3. Hence, $\mu_u / E_u = \{\{e, h, j\}, \{f, g, h\}\}$ i.e., it is a two-element quotient group, similarly for other cuts.

This Ω-group is obtained by the technique described in Theorem 3. The closure system i.e., the lattice Ω is $\mathrm{Con_w}(S_3) \setminus \Delta_{S_3}$, consisting of all weak congruences on S_3 except the diagonal Δ_{S_3}. And the order in this lattice is dual to the set inclusion.

4 Conclusion

As presented here, our paper is related to structures with generalized equality and we investigate their algebraic properties. The lattice Ω mainly generalizes equality and belonging. We do not use it to model the corresponding logic as it was the case in the basic researches by Fourman, Scott, Pavelka ([6,10]). This would be our following task. And also, in the algebraic sense, Birkhoff's variety theorem should be checked in this framework.

References

1. R. Bělohlávek, V. Vychodil, *Algebras with fuzzy equalities*, Fuzzy Sets and Systems 157, pp. 161–201, 2006. 97
2. B. Budimirović, V. Budimirović, B. Šešelja, A. Tepavčević, *E-fuzzy groups* Fuzzy Sets and Systems, 289, pp. 94–112, 2016. 97, 100, 101
3. G. Czédli, B. Šešelja, A. Tepavčević, *Semidistributive elements in lattices; application to groups and rings*, Algebra Univers. 58, pp. 349–355, 2008. 98
4. G. Czédli, M. Erné, B. Šešelja, A. Tepavčević, *Characteristic triangles of closure operators with applications in general algebra*, Algebra Univers. 62, pp. 399–418, 2009. 98
5. M. Demirci, *Foundations of fuzzy functions and vague algebra based on many-valued equivalence relations part I: fuzzy functions and their applications, part II: vague algebraic notions, part III: constructions of vague algebraic notions and vague arithmetic operations*, Int. J. General Systems, 32 (3), 123–155, 157–175, 177–201, 2003. 97
6. M.P. Fourman, D.S. Scott, *Sheaves and logic*, in: M.P. Fourman, C.J. Mulvey D.S. Scott (Eds.), Applications of Sheaves, Lecture Notes in Mathematics, vol. 753, Springer, Berlin, Heidelberg, New York, pp. 302–401, 1979. 96, 99, 100, 105
7. S. Gottwald, *Universes of fuzzy sets and axiomatizations of fuzzy set theory, Part II: Category theoretic approaches*, Studia Logica, 84(1), 23-50. pp. 1143–1174, 2006. 96
8. U. Höhle, *Fuzzy sets and sheaves. Part I: basic concepts*, Fuzzy Sets and Systems, 158(11), 2007. 96
9. T. Kuraoka, N.Y. Suzuki, *Lattice of fuzzy subalgebras in universal algebra*, Algebra universalis, 47, 223–237, 2002. 97
10. J. Pavelka, *On fuzzy logic I, II, III*, Z. Math. Logik Grundlagen math. 25, 45–52, 119–134, 447–464, 1979. 97, 105
11. B. Šešelja, A. Tepavčević, *Weak congruences in universal algebra*, Symbol Novi Sad, 2002. 98
12. B. Šešelja, A. Tepavčević, *Completion of Ordered Structures by Cuts of Fuzzy Sets, An Overview*, Fuzzy Sets and Systems, 136, 1–19, 2003. 98, 101
13. B. Šešelja, A. Tepavčević, *Fuzzy Identities*, Proc. of the 2009 IEEE International Conference on Fuzzy Systems 1660–1664. 97, 100

Poset of Unlabeled Induced Subgraphs

Scott R. Sykes

Department of Mathematics
University of West Georgia, Carrollton, GA 30118
ssykes@westga.edu

Abstract. The set of all unlabeled induced subgraphs of a finite graph G can be made into a poset by defining $H_1 \leq H_2$ if H_1 is a subgraph of H_2. This paper examines some of the connections between the graph properties and the order theoretic properties of this poset. This paper then restricts the class of subgraphs to only connected induced subgraphs and looks at some of the order theoretic properties of that poset.

1 Introduction

We will be using the following notations and conventions for the remainder of this paper. Let G be a finite graph with $V(G)$ and $E(G)$ its vertex and edge set respectively. All the subgraphs that we will be considering will be induced subgraphs. If we list a subset $H \subset G$, we mean the induced subgraph of G whose vertices are the set H. We will use the notation $US(G)$ to denote the set of all non-empty unlabeled induced subgraphs of G. We will let $K_n, K_{1,n}, P_n$ and C_n be the complete, bipartite complete, path and cycle graphs on n vertices respectively (see [3] for standard graphs and terms and [1] for lattice theoretic terms). Note that $K_3 \cong C_3$, $K_{1,1} \cong P_2$ and $K_{1,2} \cong P_3$.

In their paper, Walsh and Leach looked at the poset $US(G)$ and completely characterized for which graphs G this poset was lattice ordered. They noted that the paw (on the left of Figure 1) is not lattice-ordered in their poset since P_3 and $K_1 \cup P_2$ are both induced subgraphs that contain K_2 and $K_1 \cup K_1$ and thus $\inf\{P_3, K_1 \cup P_2\}$ does not exist in their poset. [2]

We make the following observations about $US(G)$ for any finite graph G:

1. since the graphs in $US(G)$ are unlabeled, then $H_1 = H_2$ in $US(G)$ iff there exists a graph isomorphism $\phi : H_1 \to H_2$ and the elements of $US(G)$ are $[H]$, equivalence classes of isomorphic subgraphs

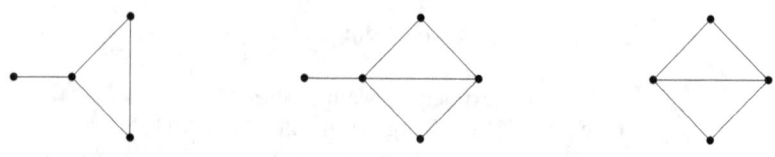

Fig. 1. Diagrams of a *paw* (left), a *dart* (center), and a *diamond* (right).

2. the 0 element in $US(G)$ is K_1 the graph consisting of a single vertex
3. if $H \prec K$ in $US(G)$, then H can be obtained by deleting a single vertex from K
4. the only possible atoms are the graphs consisting of 2 vertices, namely, the connected graph K_2 and the disconnected graph $\bar{K}_2$

Using observation 4 above, this means that if G has both at least one edge and 2 vertices that are not connected by an edge, then the poset $US(G)$ cannot be a chain since it would have K_2 and $\bar{K}_2$ as induced subgraphs. The only way a graph G can not have both of the conditions above is if either every 2 vertices of G are connected by an edge or no two vertices of G are connected by an edge. The following theorem follows directly:

Theorem 1. *For a finite graph G, the poset $US(G)$ is a chain iff $G \cong K_n$ or $G \cong \bar{K}_n$.*

2   Order Theoretic Properties

We will be making use of the following throughout this paper. If a relation R is defined on the vertex set of G by $u \ R \ v$ if $\phi(u) = v$ for some automorphism ϕ of G, then R is an equivalence relation. The equivalence classes of R are referred to as orbits, and two vertices that belong to the same orbit are called similar vertices.

Lemma 1. *Let G be a connected graph and $H \in US(G)$. If u, v are similar vertices of H, then $H \setminus \{u\} \cong H \setminus \{v\}$ in $US(G)$.*

Proof. Since u, v are similar vertices of H, there exists an automorphism of H such that $\phi(u) = v$. Since ϕ is an automorphism, $\phi : H \setminus \{u\} \to H \setminus \{v\}$. If we let xy be any edge of $H \setminus \{u\}$ so it is also an edge in H and thus G. Since ϕ is a graph automorphism, this implies that $\phi(x)\phi(y)$ is an edge of H and G as well. Since xy is an edge in $H \setminus \{u\}$, we know that $x \neq u$ and $y \neq u$ and hence that $\phi(x) \neq v$ and $\phi(y) \neq y$. Thus, $\phi(x)\phi(y)$ is an edge in $H \setminus \{v\}$ since it is an induced subgraph of G. Therefore, $H \setminus \{u\} \cong H \setminus \{v\}$ in $US(G)$. $\qquad\qquad\square$

One would hope that the above lemma would be an if and only if statement, however that is not true because of the concept of psuedo-similar vertices. Two vertices u, v in the vertex set of G are called *pseudo-similar* if $G \setminus \{u\} \cong G \setminus \{v\}$ but u and v are not similar. Let G be a graph and $V(G)$ and $E(G)$ the vertex and edge sets of G respectively. If $u, v \in V(G)$, we can define a relation by $u \mathrel{R} v$ if the graphs $G \setminus \{u\} \cong G \setminus \{v\}$. It is obvious that R is an equivalence relation on $V(G)$. The equivalence classes of this relation R we will call the *pseudo-orbits* of G.

This leads to the following useful order theoretic property:

Theorem 2. *Let G be a graph and $US(G)$ be the unlabeled induced subgraphs of G. If $H \in US(G)$ and $u, v \in V(H)$, then $H \setminus \{u\} = H \setminus \{v\}$ in $US(G)$ iff u, v are in the same pseudo-orbit of H. Thus, the number of lower covers of H in the poset $US(G)$ is the number of distinct pseudo-orbits of H.*

We would now like to define an equivalence relation on the set of vertices of G that could give us the dual of the concept of pseudo-orbits discussed above. But that proves troublesome as we explain here. For this discussion, all the sets are the induced subgraph of G with the given set as its vertices. If H_1 is a labelled subgraph of G, we can define an equivalence relation on the vertices of G by $u \mathrel{R} v$ if $H_1 \cup \{u\} \cong H_1 \cup \{v\}$. However, this paper deals with unlabeled subgraphs. So it is possible that $H_1 \cong H_2$ so that $H_1 = H_2$ as unlabeled graphs but that for some vertex u of G, $H_1 \cup \{u\} \not\cong H_2 \cup \{u\}$. So the above equivalence relation does not translate to unlabeled subgraphs. The best we can do is to define an equivalence relation on the vertex u and the graph H_i. So we are continuing to work on that.

3   Connected Subgraphs

As was pointed out above, Leach and Walsh showed that the poset for the paw graph was not lattice ordered. However, the paw graph is a connected graph and one needs a disconnected graph such as $K_1 \cup P_2$ to show it is not lattice ordered. So the question arises, if we start with a connected graph G, what happens if we restrict our poset to only the unlabeled induced subgraphs that are connected. So that is what we will explore. For the remainder of this paper, let G be a finite connected graph and we will let $CUS(G)$ be the set of all connected unlabeled induced subgraphs of G.

$CUS(G)$ is still a poset using the same order $H_1 \leq H_2$ iff H_1 is a subgraph of H_2. However, adding the condition that the subgraphs must be connected changes the poset structure of $CUS(G)$ from that of $US(G)$. So we have to find another example to show that $CUS(G)$ is not lattice-ordered. To do that, let G be the dart graph, H be the diamond and J be the paw (see Figure 1). It is obvious that both $C_3 \prec H$, $C_3 \prec J$, $P_3 \prec H$ and $P_3 \prec J$ and all 4 of these graphs are connected. Thus $\inf\{H, J\}$ does not exist in $CUS(G)$. So, $CUS(G)$ is not always lattice ordered.

When discussing the poset $US(G)$ above, we had 4 observations. The first two follow for the poset $CUS(G)$ when considering only the connected unlabeled induced subgraphs of G. However, we can update the last two as follows:

(3') For $US(G)$ one could delete any vertex. However, since we now are only looking at connected subgraphs in $CUS(G)$, one cannot delete a cut-vertex. So, if $H \prec K$ in $CUS(G)$, then H can be obtained by deleting a single vertex that is not a cut-vertex from K

(4') Since the graph $\bar{K}_2$ is disconnected, it is not in $CUS(G)$. So the only atom in $CUS(G)$ is K_2. Additionally, in $CUS(G)$ there are only 2 possible elements of height 2, namely, K_3 and P_3.

In Theorem 1, we saw that the posets $US(K_n)$ and $US(\bar{K}_n)$ are chains. Since the graph $\bar{K}_n$ is not connected, we do not have the poset $CUS(\bar{K}_n)$. It is obvious that $CUS(K_n)$ is a chain. However there are now some new graphs G that have a chain for the posets $CUS(G)$ as we will show in the following two lemmas.

Lemma 2. *For any n, $CUS(P_n)$ is a chain.*

Proof. For $n=1$ and 2, the proof is obvious. So assume $n \geq 3$. The graph P_n has 2 vertices of degree 1 and $n-2$ vertices of degree 2. However, all the vertices of degree 2 are cut vertices and the two vertices of degree

1 are similar. Thus, P_n has a unique lower cover in the poset $CUS(P_n)$, namely P_{n-1}. This implies that the poset $CUS(P_n)$ is the chain:

$$P_n \succ P_{n-1} \succ \cdots \succ P_2 \succ P_1$$

□

Lemma 3. *For any n, $CUS(C_n)$ is a chain.*

Proof. For $n=1$ and 2, the proof is obvious. So assume $n \geq 3$. All the vertices of C_n are similar so we can remove any one of them to get the unique lower cover of C_n in $CUS(C_n)$ and this unique lower cover is P_{n-1}. Using the first lemma, this implies that the poset $CUS(C_n)$ is the chain:

$$C_n \succ P_{n-1} \succ \cdots \succ P_2 \succ P_1$$

□

We can also update Theorem 2 for the poset $CUS(G)$ as follows:

Theorem 3. *Let G be a graph and $CUS(G)$ be the connected unlabeled induced subgraphs of G. If $H \in CUS(G)$ and $u, v \in V(H)$, then $H \setminus \{u\} = H \setminus \{v\}$ in $CUS(G)$ iff u, v are not cut vertices and they are in the same pseudo-orbit of H. Thus, the number of lower covers of H in the poset $US(G)$ is the number of distinct pseudo-orbits of H generated by non-cut vertices.*

References

1. G. Birkhoff. *Lattice Theory.* American Mathematical Society, Providence, RI, 3rd edition, 1967. 107
2. D. Leach and M. Walsh. A characterization of lattice-ordered graphs. *Combinatorial Number Theory,* pages 327–332, 2007. 107
3. D.B. West. *Introduction to Graph Theory.* Prentice-Hall, Englewood Cliffs, NJ, 2nd edition, 2000. 107

Author Index